I0051008

Valter Bruno Silva, João Cardoso, Antonio Chavando
Gasification

Also of Interest

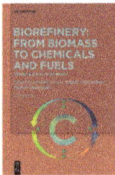

Biorefinery: From Biomass to Chemicals and Fuels.
Towards Circular Economy
Michele Aresta, Angela Dibenedetto und Franck Dumeignil (Eds.),
2022
ISBN 978-3-11-070536-2, e-ISBN 978-3-11-070538-6

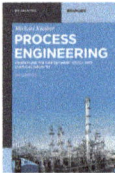

Process Engineering.
Addressing the Gap between Study and Chemical Industry
Michael Kleiber, 2020
ISBN 978-3-11-065764-7, e-ISBN 978-3-11-065768-5

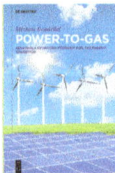

Power-to-Gas.
Renewable Hydrogen Economy for the Energy Transition
Méziane Boudellal, 2018
ISBN 978-3-11-055881-4, e-ISBN 978-3-11-055981-1

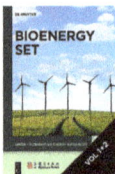

Bioenergy.
Volume 1+2
Zhenhong Yuan (Ed.), 2017
In cooperation with: China Science Publishing & Media Ltd.
Set ISBN: 978-3-11-057409-8
Volume 1: ISBN 978-3-11-034419-6, e-ISBN 978-3-11-034434-9
Volume 2: ISBN 978-3-11-047551-7, e-ISBN 978-3-11-047621-7

Valter Bruno Silva, João Cardoso,
Antonio Chavando

Gasification

—

Sustainable Decarbonization

DE GRUYTER

Authors
Dr. Valter Bruno Silva
Polytechnic Institute of Portalegre
Praça do Município 11 7300-110 Portalegre Portugal
valter.silva@ipportalegre.pt

Department of Environment and Planning/Centre for Environmental
and Marine Studies (CESAM), University of Aveiro,
Campus Universitário de Santiago, 3810-193, Aveiro, Portugal
email: v.b.silva@ua.pt

Dr. João Cardoso
Polytechnic Institute of Portalegre
Praça do Município 11 7300-110 Portalegre Portugal
jps.cardoso@ipportalegre.pt

Instituto Superior Técnico, Universidade de Lisboa, Lisboa, Portugal
Av. Rovisco Pais 1, 1049-001 Lisboa,
joaoscardoso@tecnico.lisboa.pt

Dr. Jose Antonio Mayoral Chavando
Polytechnic Institute of Portalegre Praça do Município 11 7300-110
Portalegre Portugal
antonio.chavando@ipportalegre.pt

Department of Environment and Planning/Centre for Environmental
and Marine Studies (CESAM), University of Aveiro,
Campus Universitário de Santiago, 3810-193, Aveiro, Portugal
chavando@ua.pt

ISBN 978-3-11-075820-7
e-ISBN (PDF) 978-3-11-075821-4
e-ISBN (EPUB) 978-3-11-075864-1

Library of Congress Control Number: 2022948640

Bibliographic information published by the Deutsche Nationalbibliothek
The Deutsche Nationalbibliothek lists this publication in the Deutsche Nationalbibliografie;
detailed bibliographic data are available on the internet at http://dnb.dnb.de.

© 2023 Walter de Gruyter GmbH, Berlin/Boston
Cover image: Nostal6ie/iStock/Getty Images Plus
Typesetting: Integra Software Services Pvt. Ltd.
Printing and binding: CPI books GmbH, Leck

www.degruyter.com

Acknowledgements

The authors would like to thank to the Portuguese Foundation for Science and Technology (FCT) for the grant SFRH/BD/146155/2019, contract CEECIND/00641/2018 and the projects SAICT ALT/39486/2018 and PTDC/EME-REN/4124/2021. Thanks are also due to the FCT/Ministry of Science, Technology and Higher Education (MCTES)UIDP/50017/2020+UIDB/50017/2020+LA/P/0094/2020, through national funds. This book is also a result of the project Norte-06-3559-FSE-000045 supported by NORTE 2020, under PORTUGAL 2020 Partnership agreement.

Contents

1 Gasification process and fundamentals

1.1 Gasification fundamentals

Biomass gasification is a thermochemical process in which solid biomass is transformed into valuable gases, namely CO, H_2, CH_4, and other short-chain hydrocarbons. The gasification process is partial oxidation (POX), producing O_2 and H_2O [1]. Furthermore, if the air were used as a gasifying agent, the gas mixture would also have a considerable amount of N_2 [2].

Gasification happens in steps, and each step involves a complex series of thermochemical reactions. The first step is to dry the biomass, which takes place at 100 °C. In the second stage, called pyrolysis or thermal decomposition, biochar, tar, and gas are formed. Water is lost during the pyrolysis stage (around 400 °C), and hemicellulose, cellulose, and lignin are broken down by heat. In the third stage, POX, volatiles are partially oxidized, producing heat and gases. The fourth and last stage is called "reduction." During this stage, the char and the gases react in "reduction." There are no strict borders between these phases; they often overlap and depend on the kind of gasifier, feedstock type, and process parameters like temperature, which impact the output of a gasification process incorporating the four processes mentioned earlier. The following equations describe the process of turning biomass into gas [3]:

Drying reaction

$$\text{Biomass} + \text{heat} \rightarrow \text{Biomass} + \text{water} \tag{1.1}$$

Pyrolysis reaction

$$\text{Biomass} \rightarrow H_2 + CO + CO_2 + CH_4 + C_2H_4 + C_6H_6 + C_{10}H_8 + C_6H_6O + H_2O + \text{char} \tag{1.2}$$

Oxidation reaction

$$2.5C + 2O_2 \rightarrow 1.5CO_2 \tag{1.3}$$

$$CO + 0.5O_2 \rightarrow CO_2 \tag{1.4}$$

$$CH_4 + 0.5O_2 \rightarrow CO + 2H_2 \tag{1.5}$$

$$CH_4 + 2O_2 \rightarrow CO_2 + 2H_2 \tag{1.6}$$

$$C_2H_4 + O_2 \rightarrow CO_2 + H_2O \tag{1.7}$$

$$H_2 + 0.5O_2 \rightarrow H_2O \tag{1.8}$$

$$H_2O + CO \leftrightarrow H_2 + CO_2 \tag{1.9}$$

$$C_6H_6O + 4O_2 \rightarrow 3H_2O + 6CO \tag{1.10}$$

$$C_6H_6 + 4.5O_2 \rightarrow 3H_2O + 6CO \tag{1.11}$$

https://doi.org/10.1515/9783110758214-001

$$C_{10}H_8 \rightarrow 4H_2O + 10CO \tag{1.12}$$

$$C + 2H_2 \rightarrow CH_4 \tag{1.13}$$

Reduction reaction

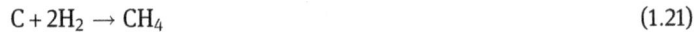

$$CO + H_2O \rightarrow H_2 + CO_2 \tag{1.14}$$

$$C + CO_2 \rightarrow 2CO \tag{1.15}$$

$$C + H_2O \rightarrow CO + H_2 \tag{1.16}$$

$$H_2O + CH_4 \rightarrow CO + 3H_2 \tag{1.17}$$

$$C_6H_6O \rightarrow CO + 0.4C_{10}H_8 + 0.15C_6H_6 + 0.1CH_4 + 0.75H_2 \tag{1.18}$$

$$C_6H_6O + 1.83333H_2O \rightarrow 2.83333CO + 0.75C_2H_4 + 1.66667CH_4 \tag{1.19}$$

$$C_6H_6 + H_2O \rightarrow 0.5CO_2 + 2C_2H_4 + 1.5C \tag{1.20}$$

$$C + 2H_2 \rightarrow CH_4 \tag{1.21}$$

The sequence of these stages varies from the type of reactor since the direction in which the feedstock and the oxidizing agent enter is different [4]. Therefore, the reactor type is one factor that influences the gasification performance parameters. These parameters are described further.

1.1.1 Gasification performance parameters

Several parameters are often quoted as a measure of performance, namely product yields, gas composition, gas lower heating values (LHVs), hot gas efficiency (HGE), cold gas efficiency (CGE), and carbon conversion efficiency (CCE) [5, 6].

1.1.1.1 Product yields
The gasification products can be classified as solid, liquid, and gas. Solid products generated from biomass gasification are not negligible. Several studies have determined alternate applications for solid chars and ashes. Typically, chars undergo secondary gasification. Other uses for char include as a precursor for activated char, fertilizers or catalysts for breakdown of NO_x (nitrous oxide) precursors, creation of synthesis gas using CO_2, and tar reforming [7]. The liquid product is also known as tar, a complex combination of chain hydrocarbons, aromatic hydrocarbons, cyclic hydrocarbons, and heterocyclic organic materials. Therefore, the composition of tar formed by pyrolysis or gasification of various organic substances is distinct [8]. Tar formation is one of the significant challenges encountered during biomass gasification. Tar can condense, causing clogged process equipment such as motors and turbines. Consequently, substantial effort has been devoted to removing tar from fuel

gas. Tar removal technologies may be roughly categorized as hot gas cleaning after the gasifier (secondary procedures) and gasifier-internal treatments (primary methods) [9]. Finally, the gas product mainly comprises CO, CO_2, H_2O, H_2, CH_4, and N_2 [10]. Several variables, including temperature, particle size, heating rate, and equivalency ratio, may influence the product yield of gasification.

The following equations outline the gasification yield estimations [11]:

$$m_F = m_{solid} + m_{gas} + m_{liquid} \tag{1.22}$$

$$Y_{solid} = \frac{m_{solid}}{m_F} \cdot 100 \tag{1.23}$$

$$Y_{gas} = \frac{m_{gas}}{m_F} \cdot 100 \tag{1.24}$$

$$Y_{liquid} = \frac{m_{liquid}}{m_F} \cdot 100 \tag{1.25}$$

where m_F represents the feedstock mass, m_{solid} is the solid mass, m_{gas} is the gas mass, m_{liquid} is the liquid mass, m_F is the feedstock mass, Y_{solid} is the solid yield, Y_{gas} is the gas yield, and Y_{liquid} is the liquid yield.

1.1.1.2 Lower heating value (LHV)

The LHV of the products is defined by the contribution of every chemical present in a particular phase. This metric is essential since it reveals the quantity of energy a product contains. The following equations determine the gas, liquid, and solid yield LHVs:

$$LHV_{gas} = \frac{\sum_{y_{igas}} \cdot m_i \cdot LHV_i}{m_{gas}} \tag{1.26}$$

$$LHV_{liquid} = \frac{\sum_{y_{iliquid}} \cdot m_i \cdot LHV_i}{m_{liquid}} \tag{1.27}$$

$$LHV_{solid} = \frac{\sum_{y_{isolid}} \cdot m_i \cdot LHV_i}{m_{solid}} \tag{1.28}$$

where y_{igas} is the mass fraction of component "i" in the gas, $y_{iliquid}$ is the mass fraction of component "i" in the liquid, y_{isolid} is the mass fraction of component "i" in the solid, m_i is the mass of component "i," m_{solid} is the solid mass, m_{gas} is the gas mass, m_{liquid} is the liquid mass, LHV_i is the LHV of the component "i," LHV_{gas} is the LHV of the gas, LHV_{liquid} is the LHV of the liquid, and LHV_{solid} is the LHV of the solid.

1.1.1.3 Cold gas efficiency (CGE)

Cold gas efficiency is the ratio of output energy to input energy [12] and may be expressed mathematically as follows:

$$CGE = \frac{LHV_{gas} \cdot m_{gas}}{LHV_F \cdot m_F} \times 100\% \qquad (1.29)$$

where CGE is the cold gas efficiency, LHV_F is the LHV of the feed stream, LHV_{gas} is the LHV of the gas mixture, m_{gas} is the mass of the gas mixture, and m_F is the mass of the feed stream.

1.1.1.4 Hot gas efficiency (HGE)

The HGE, like the CGE, is the output energy divided by the input energy but it also considers the sensible heat of the feedstock, as shown in the following equation:

$$HGE = \frac{(LHV_{Gas} \cdot m_{Gas}) + (Cp_{Gas} \cdot \Delta T \cdot m_G)}{LHV_{Feed} \cdot m_{Feed}} \times 100\% \qquad (1.30)$$

where Cp_{Gas} denotes the specific heat of produced gas, HGE indicates the hot gas efficiency, LHV_{Feed} represents the LHV of feed streams , LHV_{Gas} symbolizes the LHV of produced gas, m_{Gas} denotes the mass of the produced gas, m_F represents the mass of the feed stream, and ΔT is the temperature gradient.

1.1.1.5 Carbon conversion efficiency (CCE)

The CCE is the ratio between the mass of carbon present in the produced dry gas and the mass of carbon present in the solid fuel fed to the reactor. This ratio reflects the proportion of carbon in the solid fuel converted to carbon-containing gases that are permanent (in this case, CO_2, CO, CH_4, and C_2H_4) [13]:

$$CCE = \frac{V_{gas} \cdot \frac{P_G}{R \cdot T_G} \cdot M_C \cdot \sum_i \in C_i \cdot y_i}{m_F \cdot W_{CF}} \times 100\% \qquad (1.31)$$

where V_{gas} is the dry gas volumetric flow rate (N m³/s), P_G is the absolute pressure of the dry gas (Pa), R is the ideal gas constant (J/mol K), M_C is the molar mass of carbon (kg/mol), $\in C_i$ is the molar fraction of carbon in i (mol C/mol i), y_i is the molar fraction of CO_2, CO, CH_4, and C_2H_4 in the dry gas, m_F is the feedstock mass flow rate (kg db/s), W_{CF} is the mass fraction of carbon in the biomass (kg C/kg biomass db), and i represents the gaseous compounds such as CO_2, CO, CH_4, and C_2H_4.

In addition to the reactor type, other variables may affect the gasification performance parameters. The following section describes these parameters.

1.1.2 Parameters that influence the gasification performance

1.1.2.1 Equivalence ratio

The equivalence ratio (ER) is the mass ratio of fuel to oxidizer under stoichiometric conditions [14]. Several writers have stated that ER is the most effective method for improving gas quality [15]. In fact, its impact may be more significant than the reforming and cracking response rates [16]. The following equation represents the notion mathematically:

$$ER = \frac{m_{air}}{m_{(air)st}} \qquad (1.32)$$

where m_{air} is the mass of air corresponding to a particular ER, $m_{(air)st}$ is the mass of the air that stoichiometrically reacts with the mass of carbon and hydrogen in the feedstock, and it is calculated as follows:

$$m_{(air)st} = \left[\left(C_m \cdot \frac{MW_{O_2}}{MW_C} \right) + \left(H_{2m} \cdot \frac{0.5 MW_{O_2}}{MW_{H_2}} \right) \right] \cdot \frac{0.79}{0.21} \qquad (1.33)$$

where C_m is the mass of carbon in the feedstock, MW_C is the molecular weight of carbon, H_{2m} is the mass of hydrogen in the feedstock, MW_{H_2} is the molecular weight of hydrogen, and MW_{O_2} is the molecular weight of oxygen.

Pyrolysis conditions correlate to ER values near zero, while values equal to or greater than 1 suggests combustion conditions. Large-scale gasification commercial plants employ ER values of 0.25–0.35 because they optimize char conversion [17]. In the gasification of plastic waste, ER has a more significant impact on gas yield and reactor temperature than bed height and fluidization velocity. However, the ER quantity should be controlled since higher ER produces more CO_2, H_2O, and N_2 than CO and H_2 [15]. Nur Ashikin Jamin and coworkers [18] results are aligned with the previous statement since they concluded that when the ER rises from 0.21 to 0.37, the quantities of H_2 and CO reduce since a higher ER means more oxygen fed to the fluidized bed gasifier, increasing hydrogen oxidation (reaction (1.8)) and carbon monoxide oxidation (reaction (1.4)) gases.

Consequently, more carbon dioxide is produced at the cost of hydrogen and carbon monoxide. When ER is raised from 0.21 to 0.26, the quantity of CO_2 increases while CH_4 decreases. This situation occurs due to the oxidation of methane (reaction (1.6)) when methane reacts with oxygen from the air flow rate to form carbon dioxide and steam.

1.1.2.2 Gasification temperature

Temperature is one of the essential characteristics of the gasification process, as it affects the devolatilization process at all stages and the composition of the produced gas [19]. Therefore, gasification temperature selection is a crucial process decision [20]. Furthermore, the high volatile matter of biomass, which accelerates the solid-to-gas conversion at elevated temperatures, causes biomass conversion into gaseous products to increase as the temperature rises [21]. As lignin, a refractory component of biomass does not gasify efficiently at lower temperatures, and the thermal gasification of lignocellulosic biomass favors a minimum gasification temperature between 800 and 900 °C. However, the peak temperature of an entrained flow gasifier for biomass often exceeds 900 °C. For coal, the minimum temperature for most gasifier types is 900 °C [22].

Tar content in produced gas is dependent on gasification temperature and gasifier type. High temperature decreases the quantity of tar output (% of dry biomass mass) for a particular gasifier. The tar production is dependent on several variables, including the gasifier type and temperature [23]. Bijoy Das [24] contends that when temperature increases, tar concentration drops, H_2, CO, and CH_4 rise, and CO_2 concentration declines, enhancing the LHV of the gas. Similarly, Marek Baláš et al. [25] contend that the fluid bed temperature of a gasification reactor does not affect the final gas composition. However, the temperature on the freeboard directly affects the ultimate gas composition. High temperature increases the percentage of CO and H_2, but it reduces CH_4 and CO_2, increasing the lower calorific value of gas.

1.1.2.3 Biomass moisture

According to Atnaw et al. [26], the influence of moisture content on the H_2 and CH_4 concentrations of the produced gas was minimal, with slightly higher values obtained for fuel gasification with low moisture content. However, with a biomass moisture percentage of 29%, the calorific value of the resultant syngas was only 2.63 MJ/Nm^3, but it was almost twice (4.95 MJ/Nm^3) with a biomass moisture level of 22%. In addition, all recorded data points for the gasification of fuel with appropriate moisture content fell within the predicted range of 4–6 MJ/Nm^3 for the atmospheric gasification of biomass. This finding suggests that an initial drying of the feedstock to lower moisture content is required to create syngas with an appropriate heating value.

However, drying raw material by evaporation will consume energy, which could be greater than the energy produced in the gasification, being so an impractical process. Therefore, a possible solution is drying the biomass to the weather. Nevertheless, this option is entirely dependent on the climate [27].

1.1.2.4 Blending ratio

Co-gasification is the mixing ratio of various resources, including biomass, coal, plastic, and refuse-derived fuels (RDF). There are numerous advantages, such as the ability to achieve the desired gas composition, the improvement of gas yield and syngas quality, the reduction of tar content, the improvement of carbon conversion, cold gasification efficiency, and the enhancement of heating values for materials that are difficult to gasify using discrete processes. Co-gasification is a crucial strategy to improve the process. Still, there are additional reasons to examine the mixing ratio, such as utilizing various materials with comparable gasification performance if one material is in short supply [28, 29].

1.1.3 Gasification uses

The gasification gas is composed primarily of CO, CO_2, H_2, H_2O, N_2, CH_4, and other short-chain hydrocarbons. It is used mainly to produce electricity and heat. However, it has the potential to produce numerous chemicals like ammonia methanol and biofuels. Yet, chemical manufacturing is still in its infancy and needs much research for development and feasibility. The integration of gasification with an existing coal-fired plant at Vaskiluodon Voima Oy, Vaasa, Finland, is an excellent example of effective gasification. This integration of gasification into coal-fired facilities had several benefits, such as keeping the investment cost to about one-third of a similar-sized new biomass plant, retaining the total original coal capacity, and reducing coal consumption by 40% by using local biomasses such as wood, peat, and straw producing 230 MW of electricity and 170 MW of district heating [30].

Another example is ThyssenKrupp, whose primary product is syngas, which has various applications. Its by-products include slags, ash, and sulfur components. These leftovers may be recycled for use in road construction or the cement industry. The typical ThyssenKrupp gas composition is CO + H_2 >85 vol.%, CO_2 2–4 vol.%, and CH_4 0.1 vol.% [31].

Table 1.1 provides other instances of large-scale gasification across the globe. Several examples include materials such as municipal solid waste (MSW), plastics, and solid recovered fuels (SRF). The produced gas is utilized to generate heat and power [11].

Table 1.1: Large-scale gasification examples [11].

Country	References	Company	Technology	Feedstock	Producing
FI	[32, 33]	NSE Biofuels Oy Ltd.	Sumitomo heavy industries CFB	Wood residues	Heat 12 MWth
FI	[34]	Corenso United Ltd.	Sumitomo heavy industries CFB	Plastic waste	50 MWth
BE	[35]	Electrabe	Sumitomo heavy industries CFB	Wood residues	Heat 50 MWth

Table 1.1 (continued)

Country	References	Company	Technology	Feedstock	Producing
JP	[36]	HTW-Precon	ThyssenKrupp	MSW	–
FI	[37]	Lahti Energia Oy,	Valmet CFB	SRF	160 MW
FI	[30]	Vaskiluodon Voima Oy	Valmet CFB	Wood, peat, and straw	230 MW 170 MW heating
FI	[38]	RENUGAS	ANDRITZ Carbona BF)	Wood pellets, or chip	–
SE	[39]	GoBiGas	Valmet CFB	Wood residues	20 MW
ID	[40]	OKI Pulp & Paper	Valmet CFB	Bark and wood residues	110 MW ×2
USA	[41, 42]	Taylor Biomass Energy	Dual bed	MSW	–

1.1.4 Gasification process

The gasification process is depicted in Figure 1.1. There are two inputs to the process. First, biomass is fed by a screw feeder, which typically regulates mass flow by controlling the screw's speed and considering the screw shape. Usually, the air is the oxidizing agent, which is the second input. This mass flow may be regulated by a control valve, a flowmeter, and a temperature sensor. The gasification process includes a gas entrance to recycle the generated gas or introduce natural gas to increase the gasifier's initial temperature. By recirculating char and a portion of the generated gas, the process may become autothermal. The biomass and oxidizing agent then react inside the gasifier to generate gas. The solids are then separated from the gas using a cyclone. After the cyclone, the gas is purified by a gas cleaning system, allowing the turbine to produce high-pressure, high-temperature steam without the danger of boiler corrosion. Additionally, the gas may be utilized in other processes to manufacture ammonia and other high-value chemicals.

Temperature control is relevant in gasification since it influences product yields and composition. This parameter can be controlled in different ways, such as (1) Controlling the entry of airflow to the system, using a control valve and a temperature indicator transmitter, so if the temperature is higher than the desired temperature, the airflow will decrease by slightly closing the control valve. In contrast, if the temperature is lower than desired, the control valve will open slightly to increase the airflow. (2) The temperature can be controlled by controlling the produced

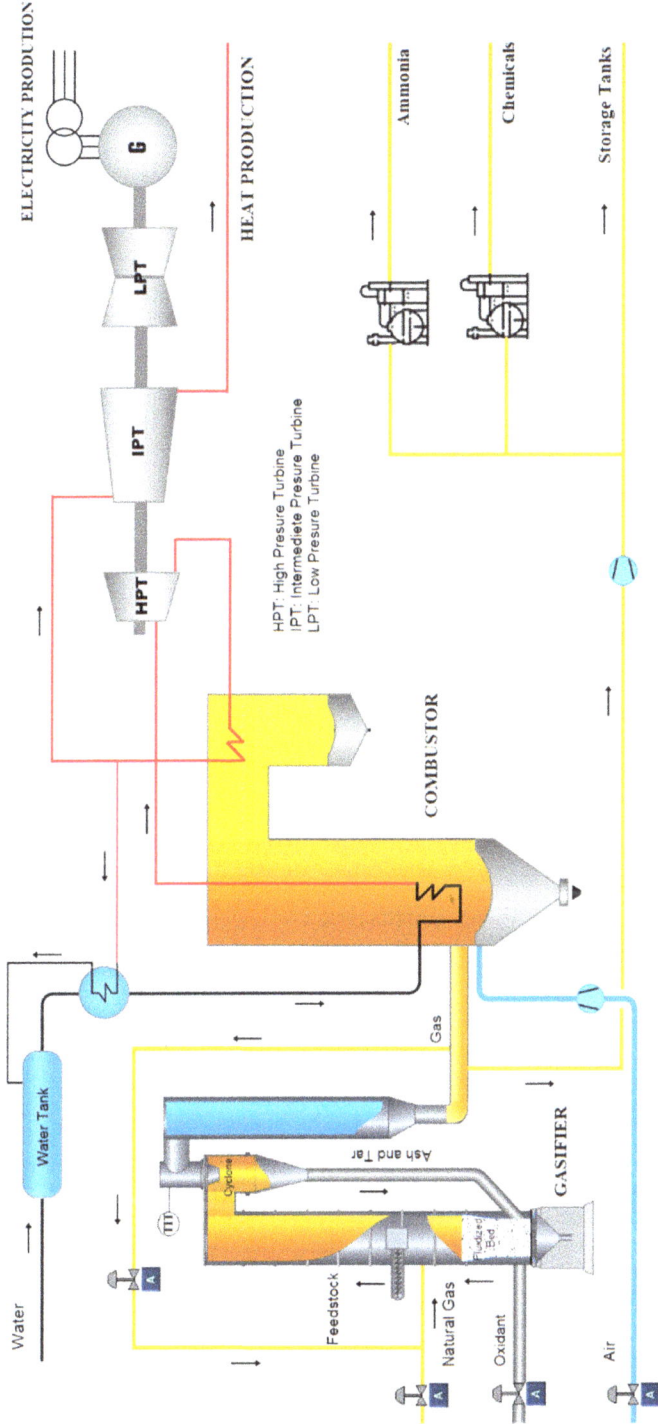

Figure 1.1: Gasification process [11].

gas recirculation following the same logic as airflow. (3) The third option considers both systems at the time, and (4) installing a serpentine inside the reactor, which contains a cold fluid, the mass flow will be controlled by following the same logic.

Gasification has several advantages, including reducing carbon dioxide emissions and eliminating fossil fuel exploitation. Another advantage is that gasification may use materials that would otherwise be discarded in landfills [43]. In addition, the electrical efficiency of waste gasification is much higher than that of direct combustion [44]. Furthermore, gasification will not produce furans as a combustion process.

Despite these benefits of the gasification process, the process is still in the optimization phase. Therefore, testing and experimentation are ongoing. Additionally, the prospect of having distinct and variable raw materials makes it more difficult to standardize the process.

1.2 History of gasification

In ancient times, when "eternal flames" were the focal point of religious sanctuaries, people were aware of flammable gases. The exterior flames were caused by underground gas sources' seepage of flammable vapors. However, a workable production technique did not emerge until much later. Numerous Europeans experimented with coal distillation, separating it into its component elements: ammonia-rich water, tar, and coke.

In 1792, William Murdock lit his home and business in Redruth, Cornwall, using coal gas [45]. Since then, gasworks, where gas was produced by thermally degrading fossil fuels and stored in gasholders, have been among the most widespread industrial complexes of the nineteenth and twentieth centuries. However, these sites declined when gaswork was replaced by natural gas [46]. Jean Tardin [47] documented in 1618 that he had heated crushed coal in a closed vessel producing coal gas after identifying the source of a fire well in Grenoble. Dr. John Clayton excavated the base of a natural spring to find coal 18 inches below. The gas escaping from the coal measures was inflammable, and Clayton assumed that the coal was the gas source. This work was unknown until Clayton published his findings in 1739 [47].

In 1760, George Dixon of Durham did tests where he heated coal in a kettle and lit the gas from the kettle's spout. In 1779, he built the first factory to get tar from coal in Cockfield, County Durham. The gas from this factory lit his house. Carlisle Spedding, the manager of Lord Lonsdale's Saltom mine in Whitehaven in 1765, lit his office with mine gas, also called "fire damp," which came out of the mine. He had offered to give the town gas for streetlights, but they had turned him down. This was done after Sir James Lowther's work in 1733 when he burned fire damp at the surface of a mine where it was being let out [45]. Archibald Cochrane, a farmer in Fife, saw that the Royal Navy was interested in coal tar, so he built a series of ovens at his family home to make it. The gas made when the coal was distilled is said to have been lit, making a bright flame that could be seen from miles away. The Royal Navy did not buy the tar, but it is thought it helped make the steam engine possible [45]. In 1785, Jean Pierre Minckelers lit the room where he taught in Louvain, France. Professor Pickel lit the fire in his chemistry lab in Wurzburg, Bavaria. In 1791, Philippe Lebon in France made gas by heating sawdust in a retort. Most of the credit for finding a way to make coal gas for sale goes to an engineer named William Murdoch [45, 48].

The first commercial gasifier was developed in 1840, and by 1878, the gasifiers worked successfully for energy production. Even J.W Parker utilized the gas generated in an automobile. From 1900 to 1945, gasification for energy generation was at its peak. However, diesel and gasoline were cheaper after WWII, so the gasification process was gradually forgotten. Furthermore, from the 1950s through the 1970s, gasification was seen as a process that was unfriendly to the medium environment since it was powered mainly by wood. Finally, in the early 1970s, scientists restarted gasification on a trial scale for waste energy production, transforming the process into an ecologically beneficial method. These days, when the indiscriminate use of fossil fuels and increasing CO_2 emissions create climate change, gasification emerges as a viable alternative for producing energy from agroforestry residues and RDF and SRF. Figure 1.2 shows the chronological order of the gasification invention.

1.3 Gasification reactors

There are a variety of types of gasifiers, which can be grouped into three types: (1) fixed bed, (2) fluidized bed, and (3) entrained flow. Their differences lie in the biomass supporting the equipment, oxidant, and feedstock flow direction, and how heat is supplied. Each gasifier type may offer distinct technological difficulties, requiring a more in-depth study outside the scope of this analysis. Table 1.2 summarizes some of their peculiarities, and Figure 1.3 shows a schematic representation of each type.

1.3.1 Updraft fixed bed (UFB)

Updraft fixed bed (UFB) gasifiers are the most straightforward and mature designs, where the oxidant flows ascendingly, and the feedstock descends [51]. The oxidant enters through a grate, which contacts the hot ashes that drop through the grate. This design has excellent tolerance to the course and feedstock moisture (60% max wet basis). However, UFB produces considerable quantities of tar 30–150 g/Nm^3, making it improper for high-volatility feedstocks [51].

1699	• Dean Clayton acquired coal gas via pyrolitic experimentation.
1760	• George Dixon of Durham heated coal in a kettle and lit the gas that came out of the kettle's spout.
1788	• Robert Gardner was awarded the first patent for gasification.
1792	• Murdoc lit a room in his house with coal gas, which was the first confirmed use of producer gas. Since then, coal gas was used for cooking and heating for a long time.
1804	• Fourcroy found water gas by combining water with a hot carbon.
1840	• France built the first commercial gasifier.
1861	• The Siemens gasifier was a real step forward in technology. This gasifier is thought to be the first one that worked well.
1878	• Gasifiers and engines worked well together to make power.
1900	• Paris showed the first 600 hp gasifier. After that, more significant engines with up to 5400 hp were used.
1901	• J.W. Parker used the produced gas in a passenger car.
1901-1920	• Many gasifier-engine systems were sold and used to make power and electricity from 1901 to 1920.
1930	• The French and British governments worked together to make charcoal gas in colonies where gasoline was hard to come by, and wood that could be burned into charcoal was easy to find and had important military and industrial uses.
After 1945	• After World War II, when gasoline and diesel were readily available and inexpensive, the gasification process lost its luster and significance.
1950-1970	• Gasification was a "Forgotten Technology" during all of these years. Many European governments thought that using wood at the rate it was being used would destroy the forests and cause environmental problems.
After 1970	• In the 1970s, people became interested again in small-scale power generation technology. Since then, people have also been trying to find ways to use other fuels besides wood and charcoal like agroforestry residues
After 2020	• Gasification uses agroforestry residuos and RDF. It is seen as potential way to procuce syngas and the producs that uses syngas like ammonia

Figure 1.2: Gasification history and development [45, 49].

1.3.2 Downdraft fixed bed (DFB)

Downdraft fixed bed (DFB) gasifiers are also concurrent. However, the produced gas goes out from the lower section, forcing the gas and tars to pass through the hot ashes, favoring cracking conditions, resulting in low tar production [55]. This design can also cope with higher moisture (25% max wet basis) and produce low amounts of tar 0.015–3.0 g/Nm^3, making it an excellent option for high-volatility feedstocks [51].

In terms of simplicity and convenience of operation, fixed bed gasifiers are very beneficial. However, they often suffer from poor mixing and heat transmission inside the gasifier, making it difficult to obtain a uniform distribution of fuel and temperature over the gasifier geometry. Furthermore, the fixed bed downdraft gasifier

Table 1.2: Gasifier types [50, 51].

Gasifier	Type	Feedstock	Oxidant	Commercialfacilities	Syngas purpose	Particle size (mm)	Tars (g/Nm³)	Heat source	Ref
Fixed bed	Updraft	Descending	Ascending	Dakota Gasification Company	Ammonia production	<51	10–150	Solid partial combustion	[52]
	Downdraft	Descending	Descending	–	–	<51	0.01–6	Volatile partial combustion	[52]
Fluidized bed	Bubbling	Ascending	Ascending	Outotec-Metso	Heat and electricity	<6	1–23	Solid and volatile partial combustion	[53]
	Circulating	Ascending	Ascending	Valmet	Heat and electricity	<6	1–30	Solid and volatile partial combustion	[54]
Entrained flow	–	Concurrent	Concurrent	ThyssenKrupp	Heat and electricity	<0.1	10–150	Solid and volatile partial combustion	[31]

has not worked well with feedstock capacities over 425 kg/h, exemplifying the technical difficulty. Furthermore, this gasifier is inappropriate for producing clean product gas (syngas) [56].

1.3.3 Bubbling fluidized bed (BFB)

In a fluidized bed reactor, gas crosses a solid granular material at high speeds to wave the solid particles and make them behave like a fluid, promoting excellent mixing and temperature homogeneity. These characteristics allow almost any feedstock type to prevent accumulation [51]. The tar production for updraft designs is ~ 50 g/Nm^3, while for downdraft designs is ~1 g/Nm^3. The oxidant can be supplied in two stages. The first is to keep the fluidization state, and the second stage is above the bed, so the entrained unreacted char particles react.

The fluidized bed gasifier works by forcing a gas stream through a particle bed vessel that acts as a fluid under certain conditions. The FBG is the most efficient type of gasifier, and its efficiency is mainly based on how the fluid and thermochemical properties of the gasifier work. So, it works best for medium-sized units between 5 and 100 MW$_{th}$ [57]. The fluidized bed gasifier is run at high pressure, which causes low volumetric gas flow rates, condensation during compression, and other problems like defluidization from particle clumping. Also, the gas that comes out of the gasification process may have a lot of particles in it, which can move around and wear down equipment. This is because crops and wastes have more ash and alkali. The alkali in the ash can combine with the silica in the sand to form eutectics that melt at a low temperature [56].

1.3.4 Circulating fluidized bed (CFB)

This technology is popular for large-scale applications due to its high efficiency [58] and suitability for fuels with high volatiles [51]. Circulating fluidized bed (CFB) circulates bed material between the cyclone and the reactor, removing ashes and returning bed material and char. This design also operates in a fluidized state, allowing temperature uniformity and intimate mixing of the hotbed material. It favors longer residence times for the gas and the fine particles than BFB reactors. Furthermore, the CFB's fluidization velocity (3.5–5.5 m/s) is higher than in BFB (0.5–1.0 m/s) [51].

1.3.5 Entrained flow (EF)

It is the most successful and widely used type of gasifier for large scale. It is typically used to gasify coal, biomass, and residues on a large scale. However, it requires pulverized fuel particles. In addition, the gasification occurs above 1000 °C, which aids in cracking tar [59]. The entrained flow gasifier (EFG) is an old alternative energy production technology used in the petroleum industry on a large scale. However, the fundamental principles underpinning its operation are still vague, particularly regarding the type of material suitable as feedstock. The EFG is operated at very high temperatures and pressures; under these operating conditions, fuel–oxygen mixtures are turned into a turbulent flame of dust. The behavior of gasification feedstock ash melting is challenging because of its high operating pressure. Due to the operating conditions, only certain feedstocks are used [56].

1.4 Gasification versus other thermochemical technologies

Biomass thermochemical conversion breaks down biomass into smaller hydrocarbon chains using heat and chemical reactions to create molecules that may be utilized to generate energy [60]. The thermochemical conversion of biomass consists mainly of three technologies [61]. Following, they are briefly described:
- **Combustion:** This technology produces heat by directly combusting biomass with excess oxygen at 800–1000 °C. This heat may be converted into mechanical and electrical energy. It is a well-known commercial technology available in residential, industrial, and utility sizes [62].
- **Pyrolysis:** It is the thermal destruction of biomass in the absence of air/oxygen. Pyrolysis of biomass is typically performed at 350–500 °C. Pyrolysis of biomass is one of the most efficient technologies used to produce biofuels – the process yields bio-oil, char, and gaseous products [62].
- **Gasification:** This process transforms biomass into valuable, convenient gaseous fuels through POX of biomass at high temperatures of 700–900 °C. Gasification of biomass also involves the removal of oxygen from the gas to increase its energy density [62, 63].

Figure 1.4 contrasts the main differences between each technology. For example, considering the input of each technology, we can see that combustion only uses biomass and air, which reacts when a heat source is applied, producing more heat and combustion gases. This technology is the simplest and most extensively studied. The primary purpose of combustion is to produce heat and electricity. On the other hand, the pyrolysis inputs are biomass and heat, which must be constantly applied to maintain temperatures ranging from 350 to 700 °C. The main objective of pyrolysis is to produce a combustible liquid (fast pyrolysis), which can be refined to produce chemicals and gasoline or burn in suitable equipment. It can also produce biochar (slow pyrolysis). Finally, biomass, air, and an initial heat source are gasification inputs. Although the difference is from combustion, the air supply is much smaller, producing mainly combustible gases which can be directly burned to

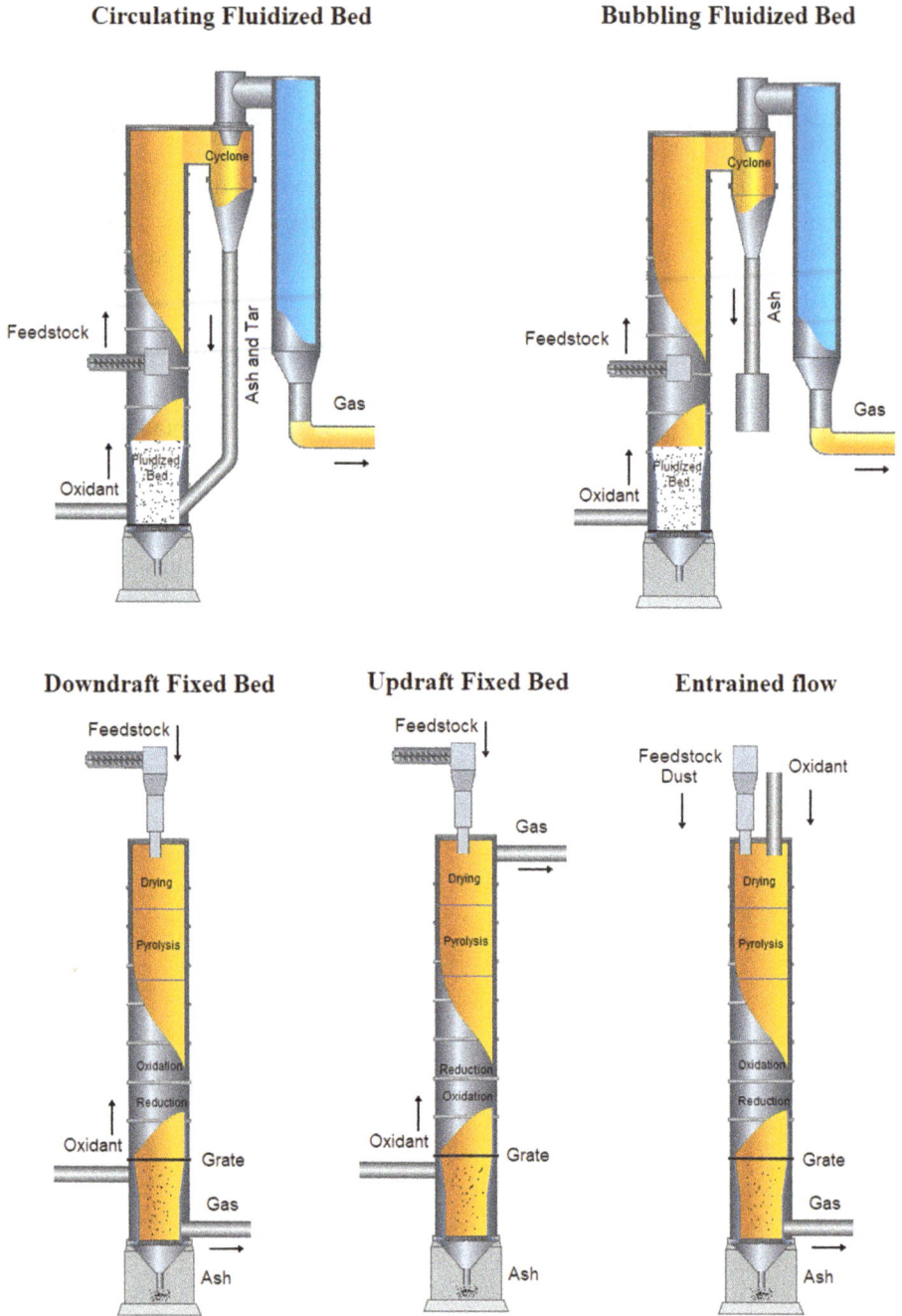

Figure 1.3: Schematic representation of gasifier types [4].

Technology | **ER** | **Temperature** | **Products** | **Process** | **Added Value Product**

Combustion — >1 — 800 - 1000 °C
Biomass + Air → Gases + Solids + Heat

Gases: CO_2, H_2O, N_2
Solids: Ashes
Heat → Electricity

Pyrolysis — 0 — 350 - 700 °C
Biomass + Heat → Liquids + Gases + Solids

Liquids: Bio-Oil, Aqueous phase — Extraction → Chemicals; Upgrading → Gasoline
Gases: CO, H_2, CH_4, C_2H_4 — Combustion → Electricity
Solids: Ashes, Biochar → Biochar

Gasification — 0.2-0.4 — 500 - 900 °C
Biomass + Air (ER = 0.2 – 0.35) → Liquids + Gases + Solids

Liquids: Tars — Synthesis → Chemicals; Ammonia; Methanol
Gases: CO_2, H_2O, N_2, CO, H_2, CH_4, C_2H_4 — Combustion → Electricity
Solids: Ashes, Biochar → Biochar

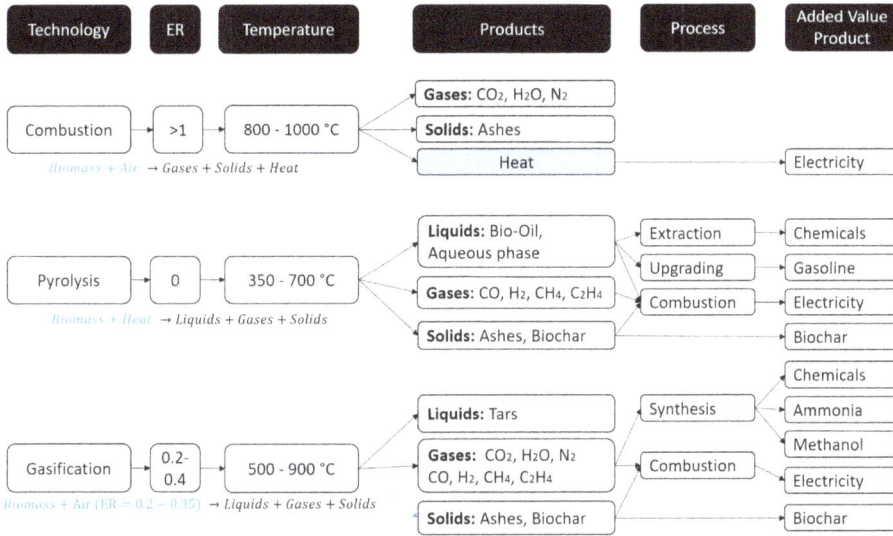

Figure 1.4: Biomass thermal conversion [61, 64].

produce heat and subsequently energy or can be refined through other processes to produce chemicals, ammonia, and methanol.

1.4.1 Combustion

Combustion burns biomass directly with excess oxygen at 800–1000 °C, where the outcome is a hot gas to be transformed into mechanical power and then electricity. It is already a well-known commercial technology, and combustion technology is broadly accessible at domestic, small industrial, and utility scales [62]. Combustion is the reaction between fuel and oxygen that makes mainly carbon dioxide and water vapor. Through a Rankine cycle, the heat given off can be used to make energy. Depending on the condition of the fuel and how it will be burned, different oven designs and starting parameters can be used to get the most out of the oven or make it work for the most extended amount of time. Direct combustion is based on a well-known technology and is currently the most popular way of obtaining energy worldwide. Modern heat and electricity cogeneration plants (coal handling plant, CHP), which also make heat for heating cities, can have a combined heat and electricity efficiency of more than 90% [65].

The various solid fuel combustion processes are determined mainly by the kind of fuel, the size, and the application (i.e., district heating, power, combined heat, and power). In any event, a combustion unit consists of four primary systems, namely (1) fuel feeding system, (2) air supply system, (3) combustion chamber, and (4) ash disposal system, which may also collect ash from the boiler [66]. In addition, there is the flue gas recirculation system, the heat exchange system (such as water tubes or shell-and-tube boilers), and the regulation (automatic control) system. CHP facilities that generate heat and electricity are typically based on the Rankine cycle with steam superheating. At constant pressure, the steam from the boiler is superheated to a greater temperature than the saturation point. Figure 1.5 depicts the process schematically [67].

Even though gasification is a reasonably well-studied process when we talk about the combustion of biomass or RDFs, this technology presents several problems. For example, the boiler's availability and *maintenance expenses* provide the most significant operational hurdles in large-scale biomass combustion facilities. Superheater tubes can clog and erode if the biomass has chlorine and fly ash. Fuel ash may interact unexpectedly, resulting in severe fouling or bed sintering issues. [68]. Slagging may occur in grate boilers, preventing fuel from flowing on the grate. Biomass often contains corrosive alkali metals, particularly potassium, and frequently chlorine. Hence, biomass combustion in PF boilers may cause superheater corrosion [69]. As a result of fuel impurities, the use of biofuels in fluidized bed boilers may lead to several operational issues. With high alkali fuels, the sintering of bed material is an issue.

Fouling of combustor surfaces is a critical issue that has substantially affected the design and operation of combustion equipment. Slagging and fouling hinder the transfer of heat and induce corrosion and erosion, which diminish the equipment's life span. Therefore, biomass may result in unscheduled downtime owing to corrosion, fouling, and slagging in high-temperature convective parts (the superheater). In addition, fouling the heat transfer surfaces results in the tubes being coated with deposits that may halt heat transfer, restrict gas passageways, accelerate the gas flow, reduce pressure, and cause structural damage due to the uneven heat flow [70].

During bed *agglomeration*, the ash particles in the fluidized bed begin agglomerating with one another and with the sand particles [71]. However, the lumps do not disappear, and the bed ceases to function within seconds. Because biomass fuels include specialized components that convert to ash, they provide a distinct challenge. For most wood fuels, coating-induced agglomeration is believed to be the dominating bed agglomeration mechanism, followed by the assault (reaction) and diffusion of calcium into the quartz, as well as the direct production of low-melting potassium silicate by potassium compounds in the biomass.

Ash melting is also another problem. Deposits on the heat transfer surfaces of a boiler are dynamic and dependent on process parameters. Typically, ash deposition is time-dependent, that is, the characteristics of ash vary throughout a deposit cross section. The most significant characteristic of ash is its strength, which is dependent on its melting behavior and sintering rate. The strength of a deposit primarily relies on its porosity [66].

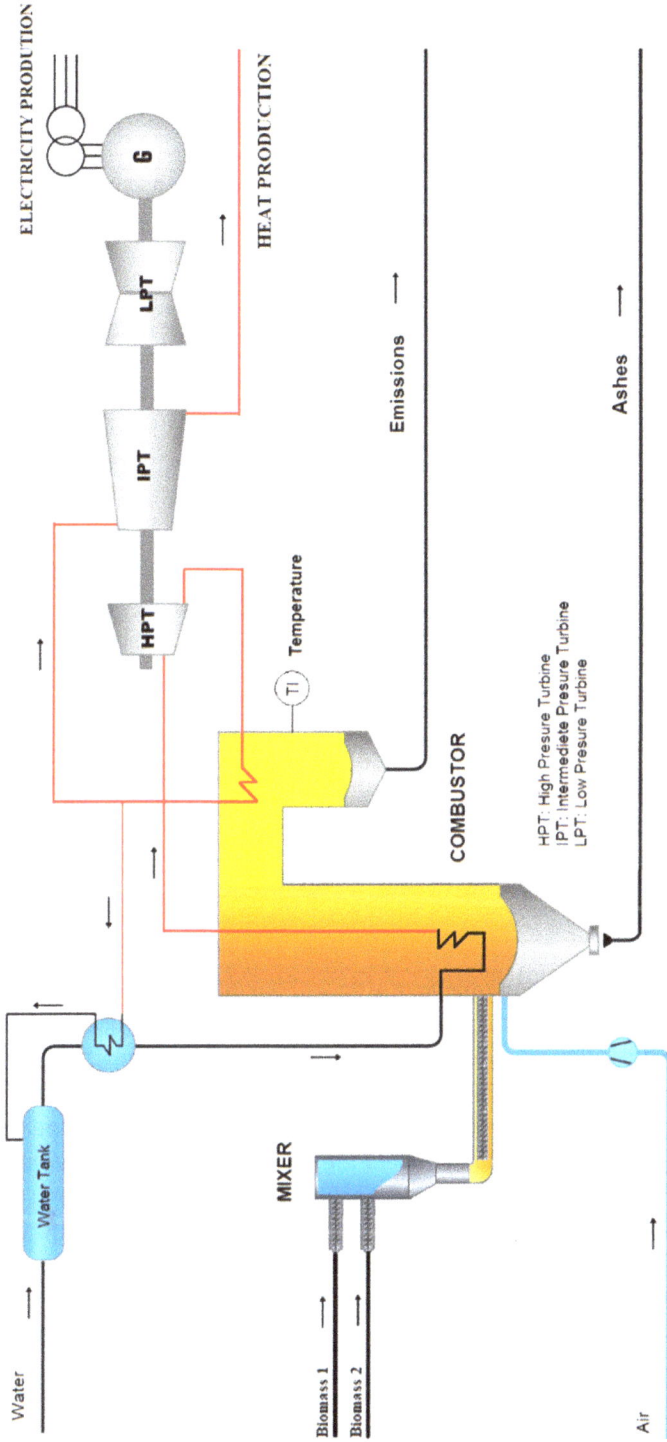

Figure 1.5: Combustion process.

Biomass combustion generates solid and *gaseous pollutant emissions* that must be treated to comply with current or future emission regulations and reduce any adverse impacts on human health and the environment. Primary pollutants from combustion may be divided into four categories: (1) complete combustion, (2) partial combustion, (3) other from inorganic species in biomass fuel, and (4) emission from N_2.

- CO_2 and H_2O are emissions coming from a complete combustion.
- CO, C_xH_y, polycyclic aromatic hydrocarbons, tar, unburned char emissions, and soot result from incomplete combustion.
- The presence of inorganic species results in the emission of particulate matter, SO_x, HCl, salts (KCl), polychlorinated dibenzo-*p*-dioxins, polychlorinated dibenzo-*p*-dibenzofurans, and heavy metals.
- The fuel's combustion conditions and nitrogen content significantly affect the formation of nitrogen oxides (NO_x, NO, NO_2, and N_2O).

We tend to postpone change until circumstances get so dire that we can no longer escape it. A long time ago, combustion was an excellent solution for energy issues. Today, however, the situation has deteriorated since pollutants from the burning of fossil fuels have produced environmental difficulties. This is the primary issue with combustion. That is why different technologies like gasification intend to help mitigate climate change. Table 1.3 describes the main differences between both technologies.

Similar to gasification, pyrolysis is a method that tries to combat climate change by generating energy from nontraditional materials such as RDF and high-density polyethylene. The section that follows covers this technique briefly and compares it with gasification.

1.4.2 Pyrolysis

Pyrolysis is the heat decomposition of organic matter without air or oxygen. Biomass pyrolysis begins at 350–500 °C and can reach 700 °C, producing liquids, gases, and solid products [62]. The pyrolysis process may be categorized as slow or rapid depending on the heating rate. In the slow pyrolysis process, the time required to heat the biomass substrate to pyrolysis temperature is greater than the retention duration of the substrate at the pyrolysis reaction temperature. In rapid pyrolysis, however, the initial heating time of the precursors is less than their ultimate retention period at the pyrolysis peak temperature. Depending on the medium, pyrolysis may also be classified as hydrous or hydropyrolysis. Slow and rapid pyrolysis is typically conducted in an inert environment [73].

Hydrous pyrolysis is conducted in the presence of water, while hydropyrolysis is conducted in the presence of hydrogen. For slow pyrolysis, the residence duration of vapor in the pyrolysis medium is prolonged. This procedure is mainly used for char manufacturing. Consequently, it may be subdivided into carbonization and

Table 1.3: Comparison between combustion and gasification [56, 72].

	Combustion	Gasification
Inputs	Biomass + air	Biomass + air
Products	Liquid (0%) Gases (95–99%) Solids (1–5%) Ashes	Liquid (2–5%) Gases (50–70%) Solids (20–40%)
ER	1–3	0.2–0.3
Temperatures	>1000 °C	700–900 °C
Main use	Heat and electricity production	Heat and electricity production
Added value products	–	Syngas, ammonia, and chemicals
Direct emissions	CO_2, NO_x, and furans	0
Main problems	High quantity of emissions Use of fossil fuels When biomass is used, it can have ash melting, agglomeration, fouling problems, and high maintenance expenses	Feedstock Homogeneous solid fuels (particle size, moisture, etc.) Process Feeding system Plugging of process parts Temperature control Gas Diluted gas with N_2 if the air is used LHV <4 MJ/m^3 Transport, not infrastructure ready

conventional. In contrast, fast pyrolysis has the residence duration of vapor for only seconds or milliseconds. This form of pyrolysis is mainly used to produce bio-oil and gas [73].

Given how important it is to act on climate change and technology's importance, pyrolysis of biomass is a great way to make energy sustainably and reduce pollution due to combustion. It has many benefits, including making high-energy products that could eventually replace nonrenewable fossil fuels [74, 75]. It is flexible and can be used with different feedstocks, such as wood and agroforestry waste [76]. Pyrolysis can also be used to break down complex materials like plastics [77], tires [78], and some parts of MSW, also called RDF [79] and SRF. Unlike burning, pyrolysis does not release dioxin and furan, which are very dangerous [80]. Also, adding char to the soil may reduce carbon more than other types of bioenergy [81]. Figure 1.6 shows the pyrolysis process integrated into a CHP in Joensuu, which was built in 2012 and went into the entire operation in 2015 [82]. It makes heat, electricity, and

50,000 tons of bio-oil (maximum planned capacity per year). The process uses a flu-idized bed boiler to heat the pyrolysis reactor and burn the coke, biochar, and non-condensed gases made during the pyrolysis process to make electricity and heat. In this way, the process of making pyrolyzed fuel can be made to work very well. Also, pyrolysis is a cost-effective way to make bio-oil to replace fossil oils when a fluidized bed boiler is added. The produced bio-oil can also be used to produce gasoline and other chemicals. However, this technology is in its infancy, and more research and test will be needed before commercial installations.

Pyrolysis is an excellent way to turn biomass into liquid fuel, and it is one of the most promising ways to use biomass for high-value purposes. In the past few years, a lot of research has been done using various experimental and simulation methods to look at how cellulose, hemicellulose, and lignin break down during py-rolysis. This has helped us understand how biomass pyrolysis works better. But biomass has a highly complex make-up, and many other parts, like inorganic com-pounds, significantly affect how biomass behaves when heated. Also, the complex interactions between the different biomass parts make it hard to learn more about how pyrolysis works. More research should be done using advanced measurement techniques and simulation methods to learn about biomass pyrolysis. Future re-search on biomass pyrolysis mechanisms should include the actual production process and real biomass samples. Also, more should be thought about the inter-actions between how the different parts of biomass behave during pyrolysis and how heat and mass move during the process. The amount of liquid oil that can be made from biomass pyrolysis is still limited, and it cannot be used as a direct re-placement for petroleum oil.

However, co-pyrolysis did show some good points in removing oxygen and im-proving oil quality. Also, the synergistic effects between biomass and feedstocks with high hydrogen content are unclear. So, for biomass pyrolysis, it is still important to figure out how to improve the synergistic effects and make better products. Catalytic fast pyrolysis (CFP) is an excellent way to make high-quality aromatic-rich liquid oil, and the key to developing CFP technology is the development of catalysts. Because of how well they work with aromatics, ZSM-5 catalysts have received a lot of attention [83]. Still, high coke and low liquid yield are two significant problems that make it hard to develop fast pyrolysis of biomass catalysts. So, more advanced catalysts should be made that are cheap, work well, last a long time, and work well with bio-mass. Also, combining co-pyrolysis and catalytic pyrolysis is something to consider if you want to get the most out of each technology [84].

Table 1.4 shows a comparison between gasification and pyrolysis. It is worth mentioning that both technologies are environmentally friendly, which in principle is the main reason for switching from typical carbon combustion to energy generation with gasification and pyrolysis. However, these processes can go further and produce fuels, such as gasoline, ethanol, and other high-value chemicals. The benefit of using

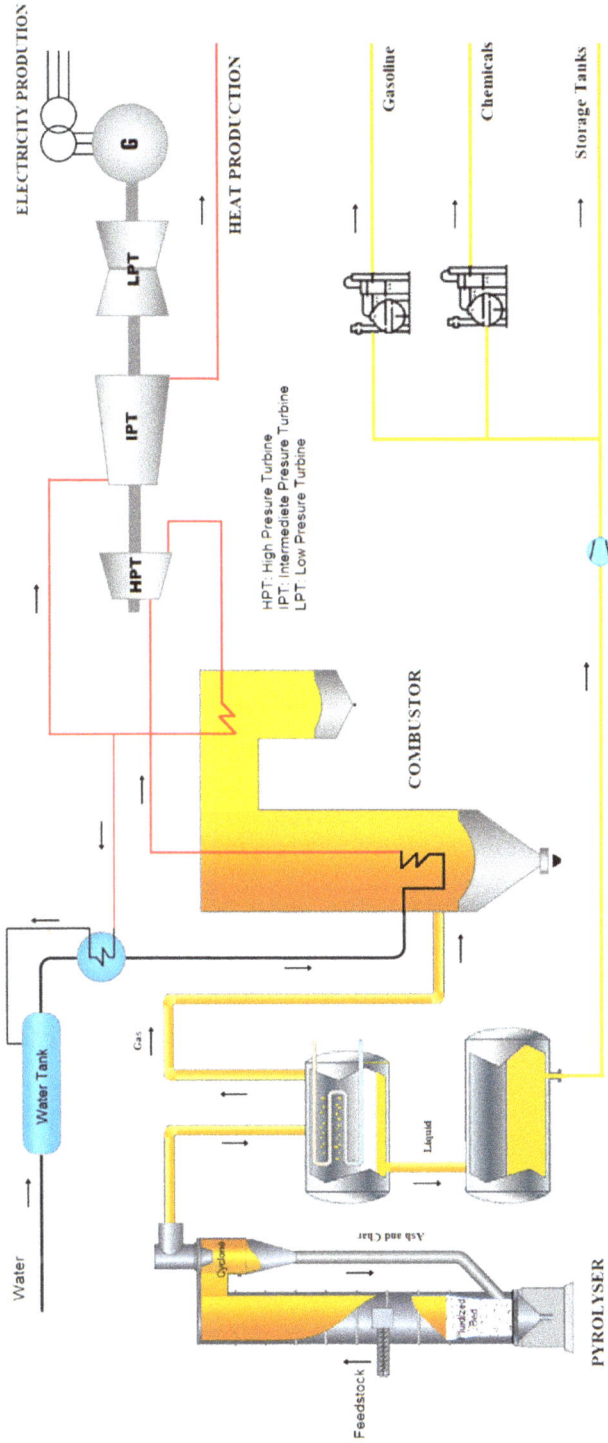

Figure 1.6: Pyrolysis process to produce energy and bio-oil [82].

Table 1.4: Comparison between combustion and gasification [56, 72, 85–87].

	Pyrolysis	Gasification
Inputs	Biomass + heat	Biomass + air
Products	Liquid (40–60%) Gases (20–30%) Solids (20–30%)	Liquid (2–5%) Gases (50–70%) Solids (20–40%)
ER	0	0.2–0.3
Temperatures	350–700 °C	700–900 °C
Main use	Heat and electricity production	Heat and electricity production
Added value products	Gasoline	Syngas, ammonia, and chemicals
Direct emissions	0	0
Main problems	Feedstock Homogeneous solid fuels (particle size, moisture, etc.) Process Feeding system Plugging of process parts Condenser, oil piping, oil tank coatings Bio-oil Acid number Particulate matter (char) Increase in pyrolysis oil viscosity over time Transport is not infrastructure ready A high number of by-products (gas and solids)	Feedstock Homogeneous solid fuels (particle size, moisture, etc.) Process Feeding system Plugging of process parts Temperature control Gas Diluted gas with N_2 if the air is used LHV <4 MJ/m^3 Transport, not infrastructure ready

these technologies to produce the mentioned products would be that the raw materials would be diverse, such as RDF, SRF, and agroforestry residues, which adds value to these products.

1.5 Future of gasification

Coal, oil, and natural gas are the principal rivals of gasification in the energy production industry. Natural gas prices are the most crucial indication of the economic viability of energy generation through gasification projects, as gasification is commercially unviable when natural gas is abundant and cheap. Current natural gas prices are USD 6.65/MMBtu (June 2022) [88], in contrast to the expenses of syngas derived by gasification, which are USD 12.3/MMBtu (for wood) and USD 5.9/MMBtu (for MSW) [89].

Despite this apparent disadvantage of gasification-produced syngas, other aspects are crucial to the success of gasification. For example, the carbon emissions futures market is a cost-effective strategy for reducing greenhouse gas emissions.

A government can limit the maximum amount of emissions generated, giving permits, or allowances, for each unit of emissions authorized under the cap to incentivize companies to reduce their emissions. Companies that emit must get and renounce emission permits. They may be granted government permissions or have economic ties with other businesses. The government may distribute the permits for free or auction them off. The EU's emissions trading scheme (ETS) is the most comprehensive system in the world for regulating emissions. It regulates emissions from over 11,000 colossal energy users, such as power plants, factories, and aviation businesses that move between ETS member states, in all 28 EU member states as well as Iceland, Liechtenstein, and Norway. It is responsible for around 45% of the EU's greenhouse gas emissions [90]. Consequently, the cost of carbon emissions futures has risen considerably, reaching a price of €85.230 (June 2022), as shown in Figure 1.7. As a result, this political approach to minimize CO_2 emissions may boost renewable energy such as gasification.

Figure 1.7: Carbon emission future cost (euros) [91].

Another crucial advantage is gasification's capacity to utilize troublesome raw materials like MSW and plastics, which have become a significant concern for humankind. Humankind created 2.01 billion tons of MSW every year [92]. Furthermore, its output is rising, with a projected 2.59 billion ton MSW generation for 2030. As a result, MSW gasification is an intriguing alternative energy-generating approach for solid waste management since it has certain benefits over traditional MSW burning

[93]. Furthermore, the produced gas can generate power and heat and chemicals like ammonia, methanol, biomethane, and liquid hydrocarbons. As a result, it is vital to underline that a waste energy recovery plant using gasification is a sanitation approach that creates energy as a beneficial by-product.

On the other hand, reliance on fossil fuels may be problematic for nations that lack the resources and must rely on other countries for consumption. As a result, novel fuels such as biomass, RDF, SRF, and agroforestry wastes are being sought. It is crucial for these countries' long-term prosperity.

References

[1]   Y. Zhang, L. Wan, J. Guan, Q. Xiong, S. Zhang, and X. Jin, "A review on biomass gasification: Effect of main parameters on char generation and reaction," *Energy and Fuels*, vol. 34, no. 11, pp. 13438–13455, Nov. 2020, doi: 10.1021/ACS.ENERGYFUELS.0C02900/ASSET/IMAGES/MEDIUM/EF0C02900_M001.GIF.

[2]   E. Shayan, V. Zare, and I. Mirzaee, "Hydrogen production from biomass gasification; A theoretical comparison of using different gasification agents," *Energy Conversion and Management*, vol. 159, pp. 30–41, Mar. 2018, doi: 10.1016/j.enconman.2017.12.096.

[3]   M. Puig-Gamero, D. T. Pio, L. A. C. Tarelho, P. Sánchez, and L. Sanchez-Silva, "Simulation of biomass gasification in bubbling fluidized bed reactor using aspen plus®," *Energy Conversion and Management*, vol. 235, pp. 113981, May. 2021, doi: 10.1016/j.enconman.2021.113981.

[4]   N. Couto, A. Rouboa, V. Silva, E. Monteiro, and K. Bouziane, "Influence of the biomass gasification processes on the final composition of syngas," *Energy Procedia*, vol. 36, pp. 596–606, Jan. 2013, doi: 10.1016/J.EGYPRO.2013.07.068.

[5]   R. Luque and J. G. Speight, *Gasification for Synthetic Fuel Production: Fundamentals, Processes and Applications*, Elsevier Ltd, 2014, doi: 10.1016/C2013-0-16368-4.

[6]   D. T. Pio, L. A. C. Tarelho, A. M. A. Tavares, M. A. A. Matos, and V. Silva, "Co-gasification of refused derived fuel and biomass in a pilot-scale bubbling fluidized bed reactor," *Energy Conversion and Management*, vol. 206, pp. 112476, Feb. 2020, doi: 10.1016/j.enconman.2020.112476.

[7]   D. A. Buentello-Montoya, X. Zhang, and J. Li, "The use of gasification solid products as catalysts for tar reforming," *Renewable and Sustainable Energy Reviews*, vol. 107, pp. 399–412, Jun. 2019, doi: 10.1016/J.RSER.2019.03.021.

[8]   R. Zeng, S. Wang, J. Cai, and C. Kuang, "A Review on Biomass Tar Formation and Catalytic Cracking," 2018.

[9]   L. Devi, K. J. Ptasinski, and F. J. J. G. Janssen, "A review of the primary measures for tar elimination in biomass gasification processes," *Biomass & Bioenergy*, vol. 24, no. 2, pp. 125–140, Feb. 2003, doi: 10.1016/S0961-9534(02)00102-2.

[10]   P. Mondal, G. S. Dang, and M. O. Garg, "Syngas production through gasification and cleanup for downstream applications – Recent developments," *Fuel Processing Technology*, vol. 92, no. 8, pp. 1395–1410, Aug. 2011, doi: 10.1016/J.FUPROC.2011.03.021.

[11]   J. A. M. Chavando, V. Silva, D. R. D. S. Guerra, D. Eusébio, J. S. Cardoso, and L. A. C. Tarelho, "Review chapter: Waste to energy through pyrolysis and gasification in Brazil and Mexico," *Gasification [Working Title]*, Jun. 2021, doi: 10.5772/INTECHOPEN.98383.

[12] P. Ponangrong and A. Chinsuwan, "An investigation of performance of a horizontal agitator gasification reactor," *Energy Procedia*, vol. 157, pp. 683–690, Jan. 2019, doi: 10.1016/j.egypro.2018.11.234.

[13] D. T. Pio, L. A. C. Tarelho, and M. A. A. Matos, "Characteristics of the gas produced during biomass direct gasification in an autothermal pilot-scale bubbling fluidized bed reactor," *Energy*, vol. 120, pp. 915–928, Feb. 2017, doi: 10.1016/J.ENERGY.2016.11.145.

[14] P. Palies. Premixed swirling flame stabilization. In *Stabilization and Dynamic of Premixed Swirling Flames*, Elsevier, 2020, pp. 105–158, doi: 10.1016/b978-0-12-819996-1.00011-1.

[15] S. A. Salaudeen, P. Arku, and A. Dutta. Gasification of plastic solid waste and competitive technologies. In *Plastics to Energy: Fuel, Chemicals, and Sustainability Implications*, Elsevier, 2018, pp. 269–293, doi: 10.1016/B978-0-12-813140-4.00010-8.

[16] R. Xiao, B. Jin, H. Zhou, Z. Zhong, and M. Zhang, "Air gasification of polypropylene plastic waste in fluidized bed gasifier," *Energy Conversion and Management*, vol. 48, no. 3, pp. 778–786, Mar. 2007, doi: 10.1016/j.enconman.2006.09.004.

[17] U. Arena, "Fluidized bed gasification," *Fluidized Bed Technologies for Near-Zero Emission Combustion and Gasification*, pp. 765–812, Jan. 2013, doi: 10.1533/9780857098801.3.765.

[18] E. M. Ranzi, R. M. Filho, N. A. Jamin, S. Saleh, A. Fazli, and A. Samad, "Influences of gasification temperature and equivalence ratio on fluidized bed gasification of raw and torrefied wood wastes," *Chemical Engineering Transactions*, vol. 80, pp. 2020, 2020, doi: 10.3303/CET2080022.

[19] M. Shahbaz, S. Yusup, A. Inayat, D. O. Patrick, M. Ammar, and A. Pratama, "Cleaner production of hydrogen and syngas from catalytic steam palm kernel shell gasification using CaO sorbent and coal bottom ash as a catalyst," *Energy and Fuels*, vol. 31, no. 12, pp. 13824–13833, Dec. 2017, doi: 10.1021/ACS.ENERGYFUELS.7B03237/ASSET/IMAGES/LARGE/EF-2017-03237Y_0006.JPEG.

[20] J. Fuchs, J. C. Schmid, S. Müller, A. M. Mauerhofer, F. Benedikt, and H. Hofbauer, "The impact of gasification temperature on the process characteristics of sorption enhanced reforming of biomass," *Biomass Conversion and Biorefinery*, vol. 10, no. 4, pp. 925–936, Dec. 2020, doi: 10.1007/s13399-019-00439-9.

[21] L. Sanchez-Silva, D. López-González, A. M. Garcia-Minguillan, and J. L. Valverde, "Pyrolysis, combustion and gasification characteristics of Nannochloropsis gaditana microalgae," *Bioresource Technology*, vol. 130, pp. 321–331, Feb. 2013, doi: 10.1016/J.BIORTECH.2012.12.002.

[22] P. Basu. Design of biomass gasifiers. In *Biomass Gasification, Pyrolysis and Torrefaction: Practical Design and Theory*, Elsevier, 2018, pp. 263–329, doi: 10.1016/B978-0-12-812992-0.00008-X.

[23] P. Basu, "Tar production and destruction," *Biomass Gasification, Pyrolysis and Torrefaction: Practical Design and Theory*, pp. 189–210, Jan. 2018, doi: 10.1016/B978-0-12-812992-0.00006-6.

[24] B. Das, *Effect of Temperature on Gasification Performance of Biomass in a Bubbling Fluidized Bed Gasifier*, 2020.

[25] M. Baláš, M. Lisý, and O. Štelcl, "The effect of temperature on the gasification process," *Acta Polytechnica*, vol. 52, no. 4, pp. 7–11, 2012, doi: 10.14311/1572.

[26] S. M. Atnaw, S. A. Sulaiman, and S. Yusup, "Influence of fuel moisture content and reactor temperature on the calorific value of Syngas resulted from gasification of oil palm fronds," *The Scientific World Journal*, vol. 2014, 2014, doi: 10.1155/2014/121908.

[27] J. A. Mayoral Chavando, "Pretreatment of wheat straw for fast pyrolysis by recirculation of by-products," Master Thesis, Lappeenranta University of Technology, Karlsruhe Institute of Technology, Lappeenranta and Karlsruhe, 2017.

[28] M. Shahbaz, T. Al-Ansari, M. Inayat, S. A. Sulaiman, P. Parthasarathy, and G. McKay, "A critical review on the influence of process parameters in catalytic co-gasification: Current performance and challenges for a future prospectus," *Renewable and Sustainable Energy Reviews*, vol. 134, pp. 110382, Dec. 2020, doi: 10.1016/J.RSER.2020.110382.

[29] J. Sousa Cardoso, V. Silva, D. Eusébio, I. Lima Azevedo, and L. A. C. Tarelho, "Techno-economic analysis of forest biomass blends gasification for small-scale power production facilities in the Azores," *Fuel*, vol. 279, pp. 118552, Nov. 2020, doi: 10.1016/J.FUEL.2020.118552.

[30] VALMET, "Fuel conversion for power boilers: Vaskiluodon Voima Oy, Vaasa, Finland," 2012. https://www.valmet.com/media/articles/all-articles/fuel-conversion-for-power-boilers-vaskiluodon-voima-oy-vaasa-finland/ (accessed Dec. 07, 2020).

[31] Thyssenkrupp, "Uhde entrained-flow gasification," 2020.

[32] Sumitomo Heavy Industries, "Biomass Gasifiers," 2020. https://www.shi-fw.com/clean-energy-solutions/biomass-gasifiers/ (accessed Dec. 07, 2020).

[33] Ltd. Sumitomo Heavy Industries, "NSE Biofuels Oy Ltd." https://www.shi-fw.com/all_projects/nse-biofuels-oy-ltd/ (accessed Dec. 07, 2020).

[34] E. Kurkela, "Review of Finnish biomass gasification technologies," 2002. https://www.researchgate.net/publication/30482338_Review_of_Finnish_biomass_gasification_technologies (accessed Dec. 07, 2020).

[35] Sumitomo, "High-value gasification solutions The power of sustainable energy solutions."

[36] M. Dobrin, "Production of Biofuels using Thyssenkrupp Gasification Technologies," 2016.

[37] VALMET, "Highest electrical efficiency from waste: Lahti Energia, Lahti Finland," 2012. https://www.valmet.com/media/articles/all-articles/highest-electrical-efficiency-from-waste-lahti-energia-lahti-finland/ (accessed Dec. 07, 2020).

[38] RENUGAS, "RENUGAS," 1993. https://www.gti.energy/renugas/ (accessed Dec. 07, 2020).

[39] VALMET, "Valmet-supplied gasification plant inaugurated at Göteborg Energi's GoBiGas in Sweden," 2014. https://www.valmet.com/energyproduction/gasification/valmet-supplied-gasification-plant-inaugurated-at-goteborg-energis-gobigas-in-sweden/ (accessed Dec. 07, 2020).

[40] VALMET, "Biomass gasification eliminates fossil fuels in the pulp mill," 2017. https://www.valmet.com/energyproduction/gasification/biomass-gasification-eliminates-fossil-fuels-in-the-pulp-mill/ (accessed Dec. 07, 2020).

[41] Taylor Biomass Energy, "The Montgomery Project," 2019. http://www.taylorbiomassenergy.com/taylorbiomass04_mont_mn.html (accessed Dec. 07, 2020).

[42] E. Voegele, "Taylor Biomass Energy project receives RES approval in New York |," Jan. 29, 2019. http://biomassmagazine.com/articles/15912/taylor-biomass-energy-project-receives-res-approval-in-new-york (accessed Dec. 07, 2020).

[43] J. A. M. Chavando, V. B. Silva, L. A. C. Tarelho, J. S. Cardoso, and D. Eusébio, "Snapshot review of refuse-derived fuels," *Utilities Policy*, vol. 74, pp. 101316, Feb. 2022, doi: 10.1016/J.JUP.2021.101316.

[44] VALMET, "Valmet Gasifier for biomass and waste," 2020. https://www.valmet.com/energyproduction/gasification/ (accessed Dec. 07, 2020).

[45] CL:AIRE, *Gasworks Profile A: The History and Operation of Gasworks (Manufactured Gas Plants) in Britain Contents*. 2014.

[46] Historic England, "Gasworks and Gasholders Introductions to Heritage Assets," 2020.

[47] J. Tardin, *Histoire Naturelle de la Fontaine Qui Brusle Près de Grenoble. Avec la Recherche de Ses Causes, & Principes, & Ample Traicté Des Feux Souterrains*, 1618.

[48] Britannica, "William Murdock." https://www.britannica.com/biography/William-Murdock-Scottish-inventor (accessed Jun. 06, 2022).

[49] NZDL.org, "Gasification History and Development." http://www.nzdl.org/cgi-bin/library.cgi?
e=d-00000-00---off-0envl--00-0----0-10-0---0---0direct-10---4-------0-1l--11-en-50---20-about---
00-0-1-00-0-0-11-1-0utfZz-8-10&cl=CL1.1&d=HASH0128ef689bc19ae2d6124941.5&gt=1
(accessed Jun. 04, 2022).

[50] A. Nuamah, A. Malmgren, G. Riley, and E. Lester, Biomass co-firing. In *Comprehensive
Renewable Energy*, vol. 5, Elsevier Ltd, pp. 55–73, 2012, doi: 10.1016/B978-0-08-087872-
0.00506-0.

[51] P. Basu. Design of biomass gasifiers. In *Biomass Gasification, Pyrolysis and Torrefaction*,
Elsevier, 2013, pp. 249–313, doi: 10.1016/b978-0-12-396488-5.00008-3.

[52] Dakota Gasification Company, "Gasification Process," 2021. https://www.dakotagas.com/
about-us/gasification/gasification-process (accessed Jun. 09, 2021).

[53] "Outotec Advanced Staged Gasifier," 2020. https://www.outotec.com/products-and-services
/technologies/energy-production/advanced-staged-gasifier/ (accessed Dec. 07, 2020).

[54] Valmet, "Valmet Gasifier for biomass and waste," 2020. https://www.valmet.com/energypro
duction/gasification/ (accessed Dec. 21, 2020).

[55] R. Muthu Dinesh Kumar and R. Anand. Production of biofuel from biomass downdraft
gasification and its applications. In *Advanced Biofuels: Applications, Technologies and
Environmental Sustainability*, Elsevier, 2019, pp. 129–151, doi: 10.1016/B978-0-08-102791-
2.00005-2.

[56] A. Mohammadi and A. Anukam, "The technical challenges of the gasification technologies
currently in use and ways of optimizing them: A review," *Energy Recovery [Working Title]*,
Feb. 2022, doi: 10.5772/INTECHOPEN.102593.

[57] P. Basu, *Biomass Gasification, Pyrolysis and Torrefaction: Practical Design and Theory*,
Elsevier Inc., 2013, doi: 10.1016/C2011-0-07564-6.

[58] J. R. Grace and C. J. Lim. Properties of circulating fluidized beds (CFB) relevant to combustion
and gasification systems. In *Fluidized Bed Technologies for Near-Zero Emission Combustion
and Gasification*, Elsevier Ltd., 2013, pp. 149–176, doi: 10.1533/9780857098801.1.147.

[59] S. Pang. Fuel flexible gas production: Biomass, coal and bio-solid wastes. In *Fuel Flexible
Energy Generation: Solid, Liquid and Gaseous Fuels*, Elsevier Inc., 2016, pp. 241–269, doi:
10.1016/B978-1-78242-378-2.00009-2.

[60] Y. H. Chan et al., "An overview of biomass thermochemical conversion technologies in
Malaysia," *Science of the Total Environment*, vol. 680, pp. 105–123, Aug. 2019, doi: 10.1016/j.
scitotenv.2019.04.211.

[61] A. V. Bridgwater, "Catalysis in thermal biomass conversion," *Applied Catalysis. A, General*,
vol. 116, no. 1–2, pp. 5–47, Sep. 1994, doi: 10.1016/0926-860X(94)80278-5.

[62] W. Y. Chen, T. Suzuki, and M. Lackner, *Handbook of Climate Change Mitigation and
Adaptation, second edition*, vol. 1–4, Springer International Publishing, 2016, doi: 10.1007/
978-3-319-14409-2.

[63] P. Basu, *Biomass Gasification, Pyrolysis and Torrefaction: Practical Design and Theory*,
Elsevier Inc., 2013, doi: 10.1016/C2011-0-07564-6.

[64] I. Y. Mohammed, Y. A. Abakr, and R. Mokaya, "Integrated biomass thermochemical
conversion for clean energy production: Process design and economic analysis," *Journal of
Environmental Chemical Engineering*, vol. 7, no. 3, Jun. 2019, doi: 10.1016/j.
jece.2019.103093.

[65] M. Mandø, "Direct combustion of biomass," *Biomass Combustion Science, Technology and
Engineering*, pp. 61–83, Jan. 2013, doi: 10.1533/9780857097439.2.61.

[66] S. Caillat and E. Vakkilainen, "Large-scale biomass combustion plants: An overview,"
Biomass Combustion Science, Technology and Engineering, pp. 189–224, Jan. 2013, doi:
10.1533/9780857097439.3.189.

[67] K. Sipilä, "Cogeneration, biomass, waste to energy and industrial waste heat for district heating," *Advanced District Heating and Cooling (DHC) Systems*, pp. 45–73, Jan. 2016, doi: 10.1016/B978-1-78242-374-4.00003-3.

[68] M. Hupa, "Ash-related issues in fluidized-bed combustion of biomasses: Recent research highlights," *Energy and Fuels*, vol. 26, no. 1, pp. 4–14, Jan. 2011, doi: 10.1021/EF201169K.

[69] D. A. Tillman, D. Duong, and B. Miller, "Chlorine in solid fuels fired in pulverized fuel boilers – sources, forms, reactions, and consequences: A literature review†," *Energy and Fuels*, vol. 23, no. 7, pp. 3379–3391, Jul. 2009, doi: 10.1021/EF801024S.

[70] T. R. Bott, "Fouling of heat exchangers," p. 524, 1995.

[71] E. Brus, M. Öhman, and A. Nordin, "Mechanisms of bed agglomeration during fluidized-bed combustion of biomass fuels," *Energy and Fuels*, vol. 19, no. 3, pp. 825–832, May. 2005, doi: 10.1021/EF0400868.

[72] J. Singh Brar, K. Singh, and J. Zondlo, "Technical Challenges and Opportunities in Cogasification of Coal and Biomass," 2012.

[73] C. Z. Zaman et al., "Pyrolysis: A sustainable way to generate energy from waste," *Pyrolysis*, Jul. 2017, doi: 10.5772/INTECHOPEN.69036.

[74] P. Ghosh, S. Sengupta, L. Singh, and A. Sahay. Life cycle assessment of waste-to-bioenergy processes: A review. In *Bioreactors*, Elsevier, 2020, pp. 105–122, doi: 10.1016/b978-0-12-821264-6.00008-5.

[75] C. Z. Zaman et al., Pyrolysis: A Sustainable Way to Generate Energy from Waste. In *Pyrolysis*, In Tech, 2017, doi: 10.5772/intechopen.69036.

[76] J. A. Ippolito et al., "Feedstock choice, pyrolysis temperature and type influence biochar characteristics: A comprehensive meta-data analysis review," *Biochar*, vol. 2, no. 4, pp. 421–438, Dec. 2020, doi: 10.1007/s42773-020-00067-x.

[77] M. S. Qureshi et al., "Pyrolysis of plastic waste: Opportunities and challenges," *Journal of Analytical and Applied Pyrolysis*, vol. 152, pp. 104804, Nov. 2020, doi: 10.1016/j.jaap.2020.104804.

[78] G. Oliveira Neto, L. Chaves, L. Pinto, J. Santana, M. Amorim, and M. Rodrigues, "Economic, environmental and social benefits of adoption of pyrolysis process of tires: a feasible and ecofriendly mode to reduce the impacts of scrap tires in Brazil," *Sustainability*, vol. 11, no. 7, pp. 2076, 2019, doi: 10.3390/su11072076.

[79] B. Dou, S. Park, S. Lim, T. U. Yu, and J. Hwang, "Pyrolysis characteristics of refuse derived fuel in a pilot-scale unit," *Energy and Fuels*, vol. 21, no. 6, pp. 3730–3734, Nov. 2007, doi: 10.1021/ef7002415.

[80] S. E. Africa, "Waste to Energy: Incineration, gasification and pyrolysis," pp. 179–180, 2019.

[81] Intergovernmental Panel on Climate Change, "Agriculture, forestry and other land use (AFOLU)," *Climate Change 2014 Mitigation of Climate Change*, pp. 811–922, 2015, doi: 10.1017/cbo9781107415416.017.

[82] IRENA, VTT, and MEAE, *Bioenergy from Finnish Forests*, 2018.

[83] Y. L. Ding, H. Q. Wang, M. Xiang, P. Yu, R. Q. Li, and Q. P. Ke, "The effect of Ni-ZSM-5 catalysts on catalytic pyrolysis and hydro-pyrolysis of biomass," *Frontiers in Chemistry*, vol. 8, Sep. 2020, doi: 10.3389/FCHEM.2020.00790.

[84] Z. Wang, "1.23 energy and air pollution," *Comprehensive Energy Systems*, vol. 1–5, pp. 909–949, Jan. 2018, doi: 10.1016/B978-0-12-809597-3.00127-9.

[85] J. A. M. Chavando, E. C. J. de Matos, V. B. Silva, L. A. C. Tarelho, and J. S. Cardoso, "Pyrolysis characteristics of RDF and HPDE blends with biomass," *International Journal of Hydrogen Energy*, Dec. 2021, doi: 10.1016/J.IJHYDENE.2021.11.062.

[86] L. Zhou, Z. Yang, D. Wei, H. Zhang, and W. Lu, "Application of Fe based composite catalyst in biomass steam gasification to produce hydrogen rich gas," *Frontiers in Chemistry*, vol. 10, pp. 372, Apr. 2022, doi: 10.3389/FCHEM.2022.882787/BIBTEX.

[87] A. Joakim, O. Anja, L. Jani, and H. Jukka, Fortum Power & Heat, and IEA Bioenergy Conference, "Challenges and opportunities with an industrial-scale integrated bio-oil plant," 2012. https://www.ieabioenergy.com/wp-content/uploads/2015/02/XII2-Autio_Challenges-and-opportunities.pdf (accessed Jun. 18, 2022).

[88] Bloomberg, "Energy," Nov. 11, 2021. https://www.bloomberg.com/energy (accessed Nov. 10, 2021).

[89] N. Indrawan, B. Simkins, A. Kumar, and R. L. Huhnke, "Economics of distributed power generation via gasification of biomass and municipal solid waste," *Energies*, vol. 13, no. 14, pp. 3703, Jul. 2020, doi: 10.3390/EN13143703.

[90] Grantham Research Institute on climate change and the environment, "How do emissions trading systems work?," Jun. 11, 2018. https://www.lse.ac.uk/granthaminstitute/explainers/how-do-emissions-trading-systems-work/ (accessed Nov. 10, 2021).

[91] Trading Economics, "EU Carbon Permits," Jun. 28, 2022. https://tradingeconomics.com/commodity/carbon (accessed Jun. 27, 2022).

[92] S. Kaza, L. Yao, P. Bhada-Tata, and F. van Woerden, *What a Waste 2.0: A Global Snapshot of Solid Waste Management to 2050*, The World Bank, 2018, doi: 10.1596/978-1-4648-1329-0.

[93] Y.-C. Seo, M. T. Alam, and W.-S. Yang, "Gasification of municipal solid waste," *Gasification for Low-grade Feedstock*, Jul. 2018, doi: 10.5772/INTECHOPEN.73685.

2 Experimental setup

2.1 Description of facilities and typical experimental conditions

The analysis performed in this book was attained from two different-sized bubbling fluidized bed biomass gasifiers: a 75 kW$_{th}$ gasifier located at the University of Aveiro [1] and a 250 kW$_{th}$ gasifier located at the Polytechnic Institute of Portalegre [2].

The installation from the University of Aveiro (Figure 2.1) integrates a pilot-scale bubbling fluidized bed reactor (75 kW$_{th}$) with a reaction chamber of 2.3 m height and 0.25 m internal diameter. About 17 kg of quartz sand, with particle size range between 355 and 1000 mm, gives the bed a static height of 0.23 m. Dry atmospheric air enters the distributor plate from the bottom of the bed. The distributor is composed of 19 injectors, each with 3 holes (1.25 mm diameter) placed perpendicularly to the direction of the gas flow in the reactor, thus providing a uniform distribution of the primary air to the bottom bed of the reactor. A preheating system heats this air stream before its injection into the gasifier vessel. Then, a screw feeder placed 0.30 m above the distributor plate discharged the biomass into the reactor.

The reactor's start-up until an operating bed temperature of around 773.1 K was done by a propane burner and preheating of the primary air. After reaching a bed temperature of around 773.1 K, the biomass feeding was started, and the gas burner and primary air preheating system were switched off. Afterward, the biomass combustion allowed the delivery of necessary heat to achieve the desired operating bed temperature. The equivalence ratio was controlled at the wanted level by adjusting the biomass feeding rate while keeping the primary air gas flow rate constant. Then, the direct gasifier was operated under autothermal and steady-state conditions without any external auxiliary heating systems. Thus, the necessary heat for the gasification process is delivered from the partial combustion of the biomass fuel in the reactor.

The fluidized bed was operated at atmospheric pressure and in a bubbling regime, with a superficial gas velocity of around 0.28–0.30 m/s (depending on the operating conditions, namely the bed temperature) and with average bed temperatures in the range of 973.15–1143.15 K. The bed temperature was maintained at the desired level by regulating the insertion of a set of eight water-cooled probes located at the bed level. In addition, nine water-cooled sampling probes placed at different heights along the reactor (two immersed in the bed and seven along the reactor freeboard) monitor the temperature and pressure inside the reactor.

Furthermore, the gasifier produced raw gas with combustible properties that supported its continuous combustion when mixed with air in an atmospheric gas burner located downstream of the gasifier; all the produced gas in the gasifier was burned continuously in the burner before being released to the atmosphere.

https://doi.org/10.1515/9783110758214-002

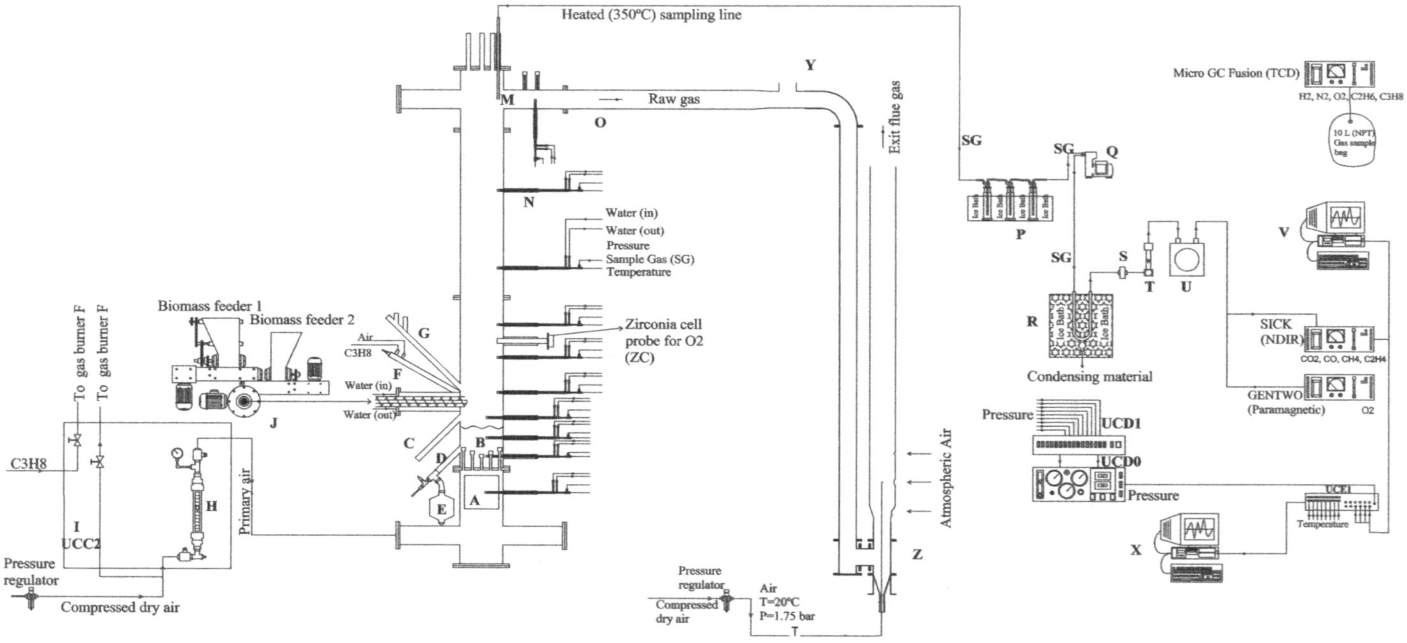

Figure 2.1: Schematic layout of the experimental gasification facility with a pilot-scale BFB reactor. Dashed line, electric circuit; continuous line, pneumatic circuit; A, primary air heating system; B, sand bed; C, bed solid level control; D, bed solid discharge; E, bed solid discharge silo; F, propane burner for preheating; G, port for visual inspection of bed surface; H, air flow meter (primary air); I, control and command unit UCC2; J, biomass feeder; M, raw gas sampling probe; N, water-cooled probe for gas sampling, pressure and temperature monitoring; O, gas exhaust; P, gas condensation unit with impingers for condensable gases (water and tars) removal; Q, gas sampling pump; R, gas condensation unit for moisture and other condensable gases removal; S, filter for particle matter/aerosol removal; T, gas flow meter; U, dry gas meter; V, computer for data acquisition from SICK analyzer; X, computer for data acquisition; Y, security exhaust pipe; Z, raw gas burner; UCD0, UCD1, electropneumatic command and gas distribution units; UCE1, electronic command unit; O_2, online gas analyzer for O_2; SICK, online gas analyzer for CO_2, CO, CH_4, and C_2H_4; micro-GC fusion, gas chromatograph with TCD [3].

The operating conditions of the reactor were characterized, namely the fuel feed rate, air feed rate, equivalence ratio, temperature and pressure along the reactor, and gas composition at the exit.

Gas samples were collected and passed through a range of detectors, including paramagnetic (O_2), nondispersive infrared (CO_2, CO, N_2O, and SO_2), chemilumines-cence (NO), and flame ionization (hydrocarbons). Further details concerning the experimental facility can be found in [4, 5].

The Polytechnic Institute of Portalegre unit consists of an upflow fluidized bed gasifier operating at up to 1123.15 K, under a total pressure below 1 bar, and at a maximum pellet feeding rate of 70 kg/h. Figure 2.2 displays a diagram of the biomass gasification plant used in the experiments.

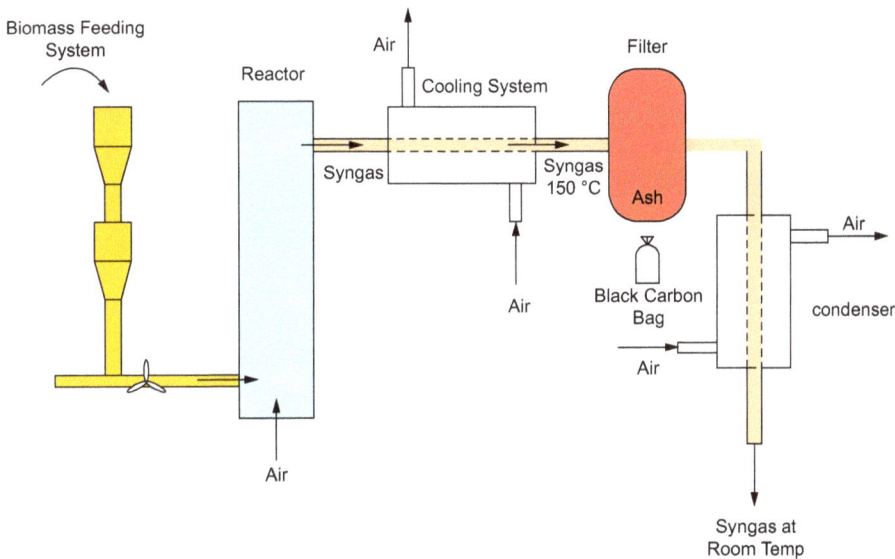

Figure 2.2: Biomass gasification semi-industrial plant at Polytechnic Institute of Portalegre, Portugal. The main components of the unit are described [2].

The main components of the unit are as follows: (a) a biomass feed system with two inline storage tanks that allow discharging the biomass into the reactor using an Archimedes' screw at variable and controllable speeds – the two storage tanks act as buffers to avoid air entering through the feeding system; (b) a tubular fluidized bed reactor, 0.5 m in diameter and 2.5 m in height, internally coated with ceramic refractory materials; biomass enters the reactor at a height of 0.4 m from the base, and preheated air enters the reactor from the base through a set of diffusers, with a flow of about 70 m^3/h (the schematic of the fluidized reactor is depicted in Figure 2.3); (c) a gas-cooling system consists of two heat exchangers: the first exchanger cools the syngas to about 570 K using a cocurrent air flow that enters the unit, and the second heat exchanger cools the syngas to 420 K by a forced flow of air coming from the exterior; (d) a cellulosic bag filter that allows the removal of carbon black and ash particles produced during the gasification process; filter cleaning is done with pressurized syngas injection; black carbon is collected at the bottom of the bag filter and stored in a proper tank; (e) a condenser where liquid condensates are removed by cooling the syngas to room temperature on a third tube heat exchanger. Gasification runs were performed using coffee husks, forest, and vine-pruning residues at 1063.15 and 1088.15 K.

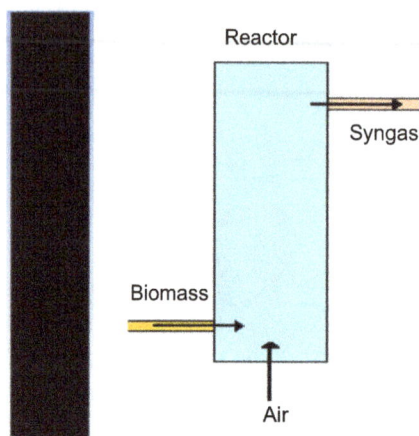

Figure 2.3: Schematic of the mesh and the corresponding bubbling fluidized bed gasifier [2].

Syngas analysis was performed in a Varian 450-GC gas chromatograph with two thermal conductivity detectors that allow the detection of H_2, CO, CO_2, CH_4, O_2, N_2, C_2H_6, and C_2H_4 (equipped, respectively, with CP81069, CP81071, CP81072, CP81073, and CP81025 Varian GC columns), using helium as carrier gas. Additional details concerning the experimental facility can be found elsewhere [6, 7].

2.2 Biomass characterization

Biomass is the only natural energy source with a lot of carbon, which can be used as a substitute for fossil fuels [8]. Part of the process of turning biomass into energy is to figure out its composition. This way, thermal analytical methods have been used extensively and fundamentally in the past few years. The promising energy production through thermochemical conversion technologies, which include pyrolysis, gasification, and combustion, depends on the thermal analysis in an actual qualification of the process [9]. Even though the elemental composition and heating value are essential for using biomass, thermal decomposition is needed to accurately figure out the properties and economic value of different samples and mixtures of biomass. The ultimate analysis or elemental analysis needs special equipment. However, the proximate analysis produces data with standard equipment like a furnace [10].

Before the actual gasification process, most biomass analyses were carried out in the Laboratory of Chemistry of the High School of Technology and Management located in Portalegre, Portugal, since biomass characteristics can provide valuable information on how the gasification process will occur. This analysis also provides crucial data to treat the implemented numerical model. The instruments used in the performed analysis are thermal gravimetric analysis (data for proximal analysis), elemental analysis (determination of biomass composition concerning the percentage of C, H, N, and O), humidity (sample moisture content assessment), and calorific value (appraisal of energy contained in biomass). As for the characterization and analysis of the Portuguese municipal solid waste (MSW) residues, these were carried out using data from the Oporto metropolitan area obtained from LIPOR, the entity responsible for the management, treatment, and recovery of MSW produced in the city. From the pretreatment of MSW conducted by LIPOR, usually via shredding and dehydration, a refuse-derived fuel containing only cellulosic and plastics is obtained [11] (the chemical composition of MSW is presented in Table 2.1). Data regarding proximate and elementary analysis for various substrates used by the authors are presented in Table 2.2.

Table 2.1: Biomass composition by income level [12].

Composition (%)	High-income	Upper-middle-income	Lower-middle-income	Low-income	Global
Food and green	32	54	53	56	44
Glass	5	4	3	1	5
Metal	6	2	2	2	4
Other	11	15	17	27	14
Paper and cardboard	25	12	12.5	7	17
Plastic	13	11	11	6.4	12
Rubber and leather	4	1	0.5	0	2
Wood	4	1	1	0.6	2

Table 2.2: Biomass properties [3, 13, 14].

Proximal analysis (%)	Balsa wood (wet basis)	Eucalyptus wood (wet basis)	Rice husks	Forest residues	Coffee husks	Miscanthus	Refuse-derived fuel (wet basis)	Pine chips	Pine pellet
Moisture (%)	6.6	11.8	10	11.3	25.3	11.4	4.3	11	4.6
Ash	1.2	2.6	18	0.2	2.5	2.1	13.4	0.3	0.3
Volatile matter	74.2	71	67	79.8	83.2	64.4	75.2	77.9	78.5
Fixed carbon	18	14.6	15	20	14.3	22.1	7.1	10.8	16.6
Elementary analysis (%, dry basis)									
Ash	–	2.87	20	–	–	–	13.4	0.3	0.3
C	50.2	45.85	39	43	40.1	44.5	54	46.4	47.5
H	6.2	6.13	5	5	5.6	5.2	7.4	6.6	47.5
N	0.1	0.35	0.4	2.4	5.2	5.3	0.5	0.2	0.1
S	–	≤100 (ppm)	0.2	–	–	–	–	–	–
O	43.5	44.80[a]	35.40[a]	49.6	49.1	45	24.1	46.5	45.9

[a]By difference.

2.2.1 Thermogravimetric characterization of biomass properties

The thermogravimetry analysis (TGA) measures how the weight of a sample changes with time or temperature in a controlled environment that could be either inert or oxidative [10]. Changes in a sample's chemistry in an oxidizing atmosphere help figure out what it is. Samples' weight changes are tracked as a temperature or time function: the heating rate or gas flow changes based on the heat used. Temperature, on the other hand, changes at a constant rate for a substance whose initial mass is known, and the changes in mass are recorded as a function of temperature over a certain amount of time. Thermography is a graph showing how mass number changes over time. A computer program controls the flow of gas into the atmosphere of some types of TGA. Different kinds of analysis can be done in each setting [10].

To determine the proximate analysis, the following standards and equations were used.

- Moisture content (M_{ad}): Biomass moisture content is crucial to its composition. It has been estimated that as the amount of moisture in a material increases, its heating value (in MJ/kg) decreases. Therefore, thermochemical conversion processes need biomass with a low moisture content to avoid a negative effect on the overall energy balance.

The moisture content is estimated by the *European Norm CEN/TS 14774–3:2004 – Solid biofuels*. It indicates to have 10 g of a biomass sample at 105 °C for 3 h and then apply the following equation:

$$M_{ad} = \frac{m_2 - m_3}{m_1} \tag{2.1}$$

where m_1 is empty crucible weight, m_2 is empty crucible weight + sample weight, m_3 is crucible weight after drying (with ashes), and M_{ad} is moisture as received.

- Ash content (A_d): Biomass ash is naturally alkaline, and its amount is usually minimal in wood. However, some biomasses can make up as much as 20% of the mass. Ashes play an essential role in biomass composition since they can produce a synergetic effect in thermochemical conversion technologies. The ash content is also estimated by the *European Norm CEN/TS 14775:2004 – Solid biofuels*. It indicates to add 4 g of sample and increase temperature at a rate of 5 °C/min until 250 °C for 60 min. Then increase the temperature at the same rate until 550 °C for 120 min:

$$A_d = \left[\left(\frac{m_6 - m_4}{m_5 - m_4} \right) 100 \right] \frac{100}{100 + M_{ad}} \tag{2.2}$$

where m_4 is empty crucible weight, m_5 is empty crucible weight + sample weight, m_6 is crucible weight after drying (with ashes), and A_d is moisture as received.

- Volatile material (V_d): The volatile matter is the gaseous phase that forms when the material breaks down due to heat. It can be categorized into light volatiles and tar (the more giant molecules that condense at ambient temperature). Many types of biomass already have a high amount of volatile matter. Because of this, biomass is easy to light. In addition, because biomass has a lot of oxygen, its volatile matter has a low LHV. Therefore, the material being processed in a thermochemical conversion and the operational conditions like temperature and heating rate significantly affect how much volatile matter is present [15].

  It is estimated by the *European Norm CEN/TS 15148:2005 – Solid biofuels*. It indicates to have 2 g at 900 °C for 7 min. Crucibles must have a cap:

$$V_d = \left(\frac{100(m_9 - m_8)}{(m_8 - m_7)} - M_{ad} \right) \frac{100}{100 - M_{ad}} \tag{2.3}$$

where m_7 is an empty crucible, m_8 is a crucible with a sample of biomass, m_9 is crucible with ashes (after the burning), V_d is % of volatile matter on a dry basis, and M_{ad} is moisture content as received.

– Fixed carbon (C_d): The fixed carbon is the part of the carbon left over after the volatile matter, moisture, and ash have been removed. It is mainly made of carbon, but small amounts of hydrogen, oxygen, nitrogen, and sulfur may be presented [16]:

$$C_d = 100 - (A_d + V_d) \tag{2.4}$$

TGA would give a fast, accurate, and efficient way to figure out how to use a biomass sample or mixture based on the vast amount of information it gives. TGA has been used with other analytical methods, which has helped us learn more about the properties of biomass. In industrial applications, it is essential to analyze the uncertainty of a sample of biomass. Also, using TGA to learn more about the properties of contaminated biomass and biomass mixtures is a must. These would eventually lead to a vital modeling tool that can figure out how to use biomass no matter what kind it is.

References

[1]   J. Cardoso, V. Silva, D. Eusébio, P. Brito, and L. Tarelho, "Improved numerical approaches to predict hydrodynamics in a pilot-scale bubbling fluidized bed biomass reactor: A numerical study with experimental validation," *Energy Conversion and Management*, vol. 156, pp. 53–67, Jan. 2018, doi: 10.1016/j.enconman.2017.11.005.
[2]   V. Silva, E. Monteiro, N. Couto, P. Brito, and A. Rouboa, "Analysis of syngas quality from Portuguese biomasses: An experimental and numerical study," *Energy and Fuels*, vol. 28, no. 9, pp. 5766–5777, Sep. 2014, doi: 10.1021/EF500570T.
[3]   D. T. Pio, L. A. C. Tarelho, A. M. A. Tavares, M. A. A. Matos, and V. Silva, "Co-gasification of refused derived fuel and biomass in a pilot-scale bubbling fluidized bed reactor," *Energy Conversion and Management*, vol. 206, pp. 112476, Feb. 2020, doi: 10.1016/j.enconman.2020.112476.
[4]   D. T. Pio, L. A. C. Tarelho, and M. A. A. Matos, "Characteristics of the gas produced during biomass direct gasification in an autothermal pilot-scale bubbling fluidized bed reactor," *Energy*, vol. 120, pp. 915–928, Feb. 2017, doi: 10.1016/J.ENERGY.2016.11.145.
[5]   M. Puig-Gamero, D. T. Pio, L. A. C. Tarelho, P. Sánchez, and L. Sanchez-Silva, "Simulation of biomass gasification in bubbling fluidized bed reactor using aspen plus®," *Energy Conversion and Management*, vol. 235, pp. 113981, May. 2021, doi: 10.1016/j.enconman.2021.113981.
[6]   V. Silva and J. Cardoso, *Computational Fluid Dynamics Applied to Waste-to-Energy Processes*, Elsevier, 2020, doi: 10.1016/c2018-0-00905-6.
[7]   V. B. Silva and A. Rouboa, "Using a two-stage equilibrium model to simulate oxygen air enriched gasification of pine biomass residues," *Fuel Processing Technology*, vol. 109, pp. 111–117, May. 2013, doi: 10.1016/J.FUPROC.2012.09.045.
[8]   M. Balat and G. Ayar, "Biomass energy in the world, use of biomass and potential trends," vol. 27, no. 10, pp. 931–940, Jul. 2006, http://dx.doi.org/10.1080/00908310490449045, doi: 10.1080/00908310490449045.

[9]   L. Zhang, C. C. Xu, and P. Champagne, "Overview of recent advances in thermo-chemical conversion of biomass," *Energy Conversion and Management*, vol. 51, no. 5, pp. 969–982, May. 2010, doi: 10.1016/J.ENCONMAN.2009.11.038.

[10] O. O. Olatunji, S. A. Akinlabi, M. P. Mashinini, S. O. Fatoba, and O. O. Ajayi, "Thermo-gravimetric characterization of biomass properties: A review," *IOP Conference Series: Materials Science and Engineering*, vol. 423, no. 1, p. 012175, Oct. 2018, doi: 10.1088/1757-899X/423/1/012175.

[11] S. Teixeira, E. Monteiro, V. Silva, and A. Rouboa, "Prospective application of municipal solid wastes for energy production in Portugal," *Energy Policy*, vol. 71, pp. 159–168, Aug. 2014, doi: 10.1016/J.ENPOL.2014.04.002.

[12] J. A. M. Chavando, V. B. Silva, L. A. C. Tarelho, J. S. Cardoso, and D. Eusébio, "Snapshot review of refuse-derived fuels," *Util Policy*, vol. 74, p. 101316, Feb. 2022, doi: 10.1016/J.JUP.2021.101316.

[13] J. Cardoso, V. Silva, D. Eusébio, P. Brito, M. J. Hall, and L. Tarelho, "Comparative scaling analysis of two different sized pilot-scale fluidized bed reactors operating with biomass substrates," *Energy*, vol. 151, pp. 520–535, May. 2018, doi: 10.1016/J.ENERGY.2018.03.090.

[14] N. D. Couto, V. B. Silva, E. Monteiro, A. Rouboa, and P. Brito, "An experimental and numerical study on the Miscanthus gasification by using a pilot scale gasifier," *Renewable Energy*, vol. 109, pp. 248–261, Aug. 2017, doi: 10.1016/J.RENENE.2017.03.028.

[15] S. Caillat and E. Vakkilainen, "Large-scale biomass combustion plants: An overview," *Biomass Combustion Science, Technology and Engineering*, pp. 189–224, Jan. 2013, doi: 10.1533/9780857097439.3.189.

[16] D. K. Sarkar, "Fuels and combustion," *Thermal Power Plant*, pp. 91–137, Jan. 2015, doi: 10.1016/B978-0-12-801575-9.00003-2.

3 Numerical modeling formulation

3.1 Equilibrium models

Thermodynamic equilibrium models are very useful tools to study the influence of most parameters for any biomass system because of their gasifier design independence [1–3]. The thermodynamic equilibrium models generally consider two approaches, giving the same results: the stoichiometric approach, which requires a clearly defined reaction mechanism that incorporates all chemical reactions and species involved; and the nonstoichiometric approach, which is based on the system minimization of the Gibbs free energy. In the last approach, the only input needed to specify the feed is its elemental composition, which can be readily obtained from ultimate analysis data [1]

The thermodynamic equilibrium approach reasonably estimates the maximum achievable yield of a certain product. On the other hand, the equilibrium approach is also independent of the gasifier design. So, it is an excellent departure point to study all the gasification potential from biomass pine residues. Furthermore, using the stoichiometric method implies incorporating all chemical reactions involving the species present in largest amounts. Under gasification conditions, it was shown that the main species present at concentrations higher than 10–4 mol% are CO, CO_2, CH_4, H_2, N_2, H_2O, and solid carbon (graphite) [4–6]. The development of the present equilibrium model is based on the following assumptions (typically assumed for equilibrium models) [6]:

1. The reactor is considered zero-dimensional.
2. The gasifier is frequently considered a perfectly insulated apparatus, that is, heat losses are neglected. In practice, gasifiers have heat losses to the environment, but this term can be incorporated into the enthalpy balance of the equilibrium model.
3. The dependence on the hydrodynamic behavior as a function of the reactor design is considered perfect mixing and uniform temperature. This last assumption is well suited for downdraft gasifiers. Indeed, the developed model is used considering a downdraft gasifier.
4. Gasification reaction rates are fast enough, and residence time is sufficiently long to reach the equilibrium state.
5. Tar content is considered negligible.

The equilibrium model is subdivided into two different stages:

1. Heterogeneous equilibrium, where all the compositions are found below and at the carbon boundary point (CBP). At this stage, there is the presence of unconverted solid carbon. Its total depletion occurs only at the CBP.
2. Homogeneous equilibrium, where all the compositions are in the gaseous state without any unconverted solid carbon.

https://doi.org/10.1515/9783110758214-003

3.1.1 Heterogeneous equilibrium

3.1.1.1 Mass balance

The global gasification reaction can be written as follows:

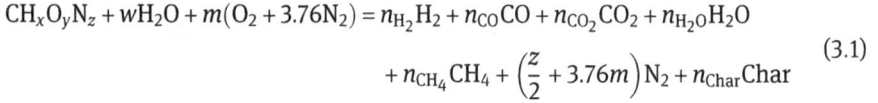

$$CH_xO_yN_z + wH_2O + m(O_2 + 3.76N_2) = n_{H_2}H_2 + n_{CO}CO + n_{CO_2}CO_2 + n_{H_2O}H_2O$$

$$+ n_{CH_4}CH_4 + \left(\frac{z}{2} + 3.76m\right)N_2 + n_{Char}Char \tag{3.1}$$

where x, y, and z are the number of atoms of hydrogen, oxygen, and nitrogen per number of carbon atoms in the feedstock, and w and m are the amount of moisture and oxygen per kmol of feedstock, respectively. The subscripts x, y, and z are determined from the ultimate analysis of the biomass. The proximate and ultimate analyses of the biomass pine residues (minor components such as S or Cl are not included in Table 3.1) are given in Table 3.1.

Table 3.1: Proximate and ultimate analyses of biomass pine residues.

	Pine residues
Proximate analysis, %	
Ash	0.41
Moisture	9.80
Volatile matter	73.61
Fixed carbon	15.58
Ultimate analysis ($CH_xO_yN_z$)	
x	1.510
y	0.610
z	0.0017

Regarding the global gasification reaction, the atomic balance of carbon (eq. (3.2)), hydrogen (eq. (3.3)), and oxygen (eq. (3.4)) can be defined as follows:

$$n_{CO} + n_{CO_2} + n_{CH_4} + n_{Char} - 1 = 0 \tag{3.2}$$

$$2n_{H_2} + 2n_{H_2O} + 4n_{CH_4} - x - 2w = 0 \tag{3.3}$$

$$n_{CO} + 2n_{CO_2} + n_{H_2O} - w - 2m - y = 0 \tag{3.4}$$

3.1.1.2 Thermodynamic equilibrium

During the gasification process, the relevant chemical reactions for the heterogeneous equilibrium are the exothermic methane formation (eq. (3.5)), the endothermic water–gas (eq. (3.6)), and Boudouard reactions (eq. (3.7)) [7]:

$$C + 2H_2 = CH_4 \tag{3.5}$$

$$C + H_2O = CO + H_2 \tag{3.6}$$

$$C + CO_2 = 2CO \tag{3.7}$$

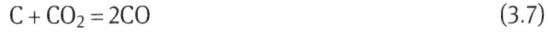

The equilibrium constants for the exothermic methane formation, the endothermic water–gas, and Boudouard reactions are defined as follows, respectively:

$$K_{eq.\,(3.5)}(T) = \frac{\left(\frac{n_{CH_4}}{n_T}\right)}{\left(\frac{n_{H_2}}{n_T}\right)^2} \left(\frac{P_{Ref}}{P}\right) \tag{3.8}$$

$$K_{eq.\,(3.6)}(T) = \frac{\left(\frac{n_{CO}}{n_T}\right)\left(\frac{n_{H_2}}{n_T}\right)}{\left(\frac{n_{H_2O}}{n_T}\right)} \left(\frac{P}{P_{Ref}}\right) \tag{3.9}$$

$$K_{eq.\,(3.7)}(T) = \frac{\left(\frac{n_{CO}}{n_T}\right)^2}{\left(\frac{n_{CO_2}}{n_T}\right)} \left(\frac{P}{P_{Ref}}\right) \tag{3.10}$$

where n_T is the sum of the species present in the product gas (eq. (3.11)), P_{Ref} is the standard pressure at 1 atm, P is the pressure at the operating condition, and $k_{eqi}(T)$ is the equilibrium constant as a function of the temperature:

$$n_T = n_{CO} + n_{CO_2} + n_{CH_4} + n_{H_2} + n_{H_2O} + n_{N_2} \tag{3.11}$$

The equilibrium constant can be obtained by the following equation:

$$k_{eqi}(T) = \exp\left(\frac{\sum_i \vartheta_i g_i^0}{RT}\right) \tag{3.12}$$

where R is the universal gas constant, 8.314 J/kmol·K, and g_i^0 is the molar-specific Gibbs function at the operating temperature and can be defined as follows:

$$g_i^0 = g_f^0 + \Delta h_{r,i} - \Delta S_{T,i} \tag{3.13}$$

where g_f^0 is the molar-specific Gibbs function of formation at the reference state, $\Delta h_{T,i}$ is the enthalpy reaction change between the operating temperature and the reference temperature, and $\Delta S_{T,i}$ is the entropy reaction change between the operating temperature and the reference temperature.

The enthalpy change (eq. (3.14)) and entropy change (eq. (3.15)) can be computed by the following equations:

$$\Delta h_{r,i} = \int_{T_{\text{Ref}}}^{T} C_{pi} dT \tag{3.14}$$

$$\Delta S_{T,i} = \int_{T_{\text{Ref}}}^{T} C_{pi} dT - R \ln \frac{P}{P_{\text{Ref}}} \tag{3.15}$$

where C_{pi} is the specific heat at constant pressure and is a function of the temperature. The enthalpy, entropy, and the ratio between the Gibbs function of formation and the entropy of formation are shown in Table 3.2 [7, 8]. C_{pi} for the gaseous species can be computed by the following empirical equation:

$$C_{pi} = a + bT + cT^2 + dT^3 \tag{3.16}$$

Table 3.2: Molar enthalpy, entropy, and Gibbs function of formation at the reference state for the chemical elements considered in the present model [7, 8].

Chemical element	S_f^0 (kJ/kmol K)	g_f^0 (kJ/kmol)	h_T^0 (kJ/mol)
CO	197.653	−137.150	−110.5
CO_2	213.795	−394.374	−393.5
CH_4	182.256	−50.751	−74.8
C(s)	5.740	0.000	0.0
H_2	130.684	0.000	0.0
N_2	191.610	0.000	0.0
O_2	205.138	0.000	0.0
H_2O (g)	188.833	−228.583	−241.8
H_2O (l)	70.049	−237.178	−285.9

where $a, b, c,$ and d are the specific gas species coefficients. These parameters can be obtained in [8, 9]. The heat capacity of solid carbon (in kJ/kmol·K) can be defined by the following polynomial equation [10]:

$$C_{p,\text{carbon}} = 17.166 + 4.271 \frac{T}{100} - \frac{8.79 \times 10^5}{T^2} \tag{3.17}$$

3.1.1.3 Energy balance
The global energy balance equation below and at the CBP can be defined as follows for 1 kg of biomass:

$$Q + \text{LHV}_{\text{biomass}} + \Delta h_{\text{water}} \cdot n_{\text{water}} = \left(\Delta h_{\text{gas}} + \text{LHV}_{\text{gas}}\right) \cdot n_{\text{gas}} + \left(\Delta h_{\text{carbon}} \cdot C_{p,\text{carbon}}\right) n_c \text{Char} \tag{3.18}$$

where Q is the heat addition to the gasification process, $LHV_{biomass}$ is the lower heating value of the biomass, $\Delta h_{air} \times n_{air}$ is the product between the air enthalpy difference at any given T and at 298 K by the molar amount of air, $\Delta h_{water} \times n_{water}$ is the product between the water enthalpy difference at any given T and at 298 K by the molar amount of water vapor, Δh_{gas} is the gas enthalpy difference at any given T and at 298 K, LHV_{gas} is the lower heating value of the gas, n_{gas} is the molar amount of the gas, $C_{p, carbon}$ is the molar-specific heat capacity of solid carbon, and n_cChar is the molar amount of carbon (char).

The high heating value (HHV) of the fuel in MJ/kg [11] and the LHV of the gas in kJ/kmol [12] can be obtained by the following equations, respectively:

$$HHV_{biomass} = 0.3941w_C + 1.1783w_H - 0.1034w_O - 0.0151w_N + 0.1005w_S - 0.0211w_A \tag{3.19}$$

Additionally, the LHV of the fuel can be computed by subtracting the HHV by the enthalpy of vaporization of water of the dry solid fuel:

$$LHV_{gas} = 282993n_{CO} + 802303n_{CH_4} + 241827n_{H_2} \tag{3.20}$$

where $w_C, w_H, w_O, w_N, w_S,$ and w_A are the respective weight fractions for carbon, hydrogen, oxygen, nitrogen, sulfur, and ash.

3.1.2 Homogeneous equilibrium

3.1.2.1 Mass balance

The balance for the N_2, C, H_2, and O_2 species at the homogeneous equilibrium can be given by eqs. (3.21), (3.22), (3.23) and (3.24), respectively:

$$\left(n_{gas} \cdot n_{N_2, \text{ at CBP}} \right) \left(n_{N_2, \text{ air}} \cdot n_{air} \right) = n_{gas} \cdot n_{N_2} \tag{3.21}$$

$$n_{gas} \left(n_{CO} + n_{CO_2} + n_{CH_4} \right)_{\text{at CBP}} = n_{gas} \left(n_{CO} + n_{CO_2} + n_{CH_4} \right) \tag{3.22}$$

$$n_{gas} \left(n_{H_2} + n_{H_2O} + n_{CH_4} \right)_{\text{at CBP}} + n_{H_2O} = n_{gas} \left(n_{H_2} + n_{H_2O} + 2n_{CH_4} \right) \tag{3.23}$$

$$n_{gas} \left(0.5 \left(n_{CO} + n_{H_2O} \right) + n_{CO_2} \right)_{\text{at CBP}} + n_{O_2, \text{ air}} {}^* n_{air} + 0.5n_{H_2O} = n_{gas} \left(0.5 \left(n_{CO} + n_{H_2O} \right) + n_{CO_2} \right) \tag{3.24}$$

where n_{gas} is the molar amount of gas, $n_{N_2;air}$ is the molar amount of nitrogen in the air, n_{CO} is the molar amount of carbon monoxide, n_{CO_2} is the molar amount of carbon dioxide, n_{CH_4} is the molar amount of methane, n_{H_2} is the molar amount of hydrogen, n_{H_2O} is the molar amount of water, $n_{O_2;air}$ is the molar amount of oxygen on the air, n_{air} is the molar amount of air, and the subscript CBP indicates the molar amount of the specie at the carbon boundary point.

3.1.2.2 Thermodynamic equilibrium

The relevant chemical reactions for the homogeneous equilibrium are the water–gas shift reaction (eq. (3.25)) and the methane formation reaction (eq. (3.26)):

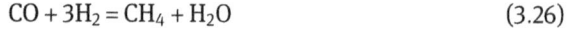

$$CO + H_2O = CO_2 + H_2O \tag{3.25}$$

$$CO + 3H_2 = CH_4 + H_2O \tag{3.26}$$

The equilibrium constants for the water–gas shift reaction and the methane formation reaction are defined as follows, respectively:

$$K_{eq. (3.25)} = \frac{\left(\frac{n_{CO_2}}{n_T}\right) \cdot \left(\frac{n_{H_2}}{n_T}\right)}{\left(\frac{n_{CO}}{n_T}\right) \cdot \left(\frac{n_{H_2O}}{n_T}\right)} \tag{3.27}$$

$$K_{eq. (3.25)} = \frac{\left(\frac{n_{CH_4}}{n_T}\right) \cdot \left(\frac{n_{H_2O}}{n_T}\right)}{\left(\frac{n_{CO}}{n_T}\right) \cdot \left(\frac{n_{H_2}}{n_T}\right)^3} \left(\frac{P_{ref}}{P}\right)^2 \tag{3.28}$$

3.1.2.3 Energy balance

The global energy balance equation above the CBP can be defined as follows:

$$\left(\Delta h_{gas} + LHV_{gas}\right)_{CBP} \cdot n_{gas, \, CBP} + \Delta h_{air} + n_{air} + \Delta h_{water} + n_{water}$$
$$+ Q_{homogeneous} = \left(\Delta h_{gas} + LHV_{gas}\right) \cdot n_{gas} \tag{3.29}$$

where $Q_{homogeneous}$ is the heat addition to the gasification process above the CBP.

3.1.3 Calculation procedure

The methodology suggested by Jarungthammachote and Dutta [7] to correct the equilibrium constants using multiplicative factors was used in this model.

To solve the values of n_{CO}, n_{CO_2}, n_{H_2}, n_{CH_4}, n_{H_2O}, n_T, T (temperature at the CBP), and Q, an initial temperature was assumed, and eqs. (3.5)–(3.11) and (3.18) were solved using the Matlab program. Equations (3.12)–(3.17) were integrated into eqs. (3.8)–(3.10). The amount of unconverted carbon is a parameter only for the heterogeneous equilibrium. The heterogeneous equilibrium means that there is the presence of unconverted solid carbon, and then the operation is being developed below or at the CBP.

The same methodology was applied to solve the homogeneous equilibrium (where there is no unconverted solid carbon) being the computed values of n_{CO}, n_{CO_2}, n_{H_2}, n_{CH_4}, and n_{H_2O} obtained from the heterogeneous equilibrium used as input parameters in this part of the model.

3.2 Computational fluid dynamics (CFD)

The exponential growth in computational power is gradually replacing empirical or semiempirical models for computational fluid dynamics (CFD) to predict biomass and waste gasification. CFD models provide crucial insights into the flow field inside the reactor, leading to a better understanding and improved operational performance while indicating solutions to potential problems. However, given the extreme complexity of creating a realistic model, this application is still in a developing stage, and more studies are required [13].

The gasification process is characterized by a multiphase flow containing solid and gas phases, in which slagging gasifier liquids can also be present. The solid particles may hold a wide range of sizes and shapes, especially when solid wastes are considered, while their organic components are consumed as they pass through the reactor [14]. The interplay between both phases is of utmost importance to model correctly since these exchange heat by convection, mass over the heterogeneous chemical reactions, and momentum due to the drag between the gas and solid phases [15]. In most cases, this becomes a highly complex process since the user must dispose of a thorough knowledge of all relevant phenomena involved (i.e., mass transfer rates, solid properties, heat transfer rates, reaction rates, the equation of state data and gas viscosities, among other features) which unfortunately are seldom available. Therefore, modeling these phenomena, beyond other required steps to create a realistic model capable of accurately predicting the gasification process, has proved to be a daunting process for most researchers.

To point out this ability, Figure 3.1 summarizes a previous analysis developed by the author's research group in which a 2D Eulerian–Eulerian model approach is set to describe the fluidization process in a 250 kW$_{th}$ fluidized bed reactor [16].

3.2.1 2D Eulerian–Eulerian approach

In the comprehensive 2D numerical model, the gas phase is treated as continuous, and the solid phase is described through an Eulerian granular model. Interactions between both phases are modeled since both phases exchange heat by convection, momentum (given the drag between the gas phase and solid phase), and mass (given the heterogeneous chemical reactions).

3.2.1.1 Mass balance model
The gas- and solid-phase continuity equations are given in eqs. (3.30) and (3.31), respectively:

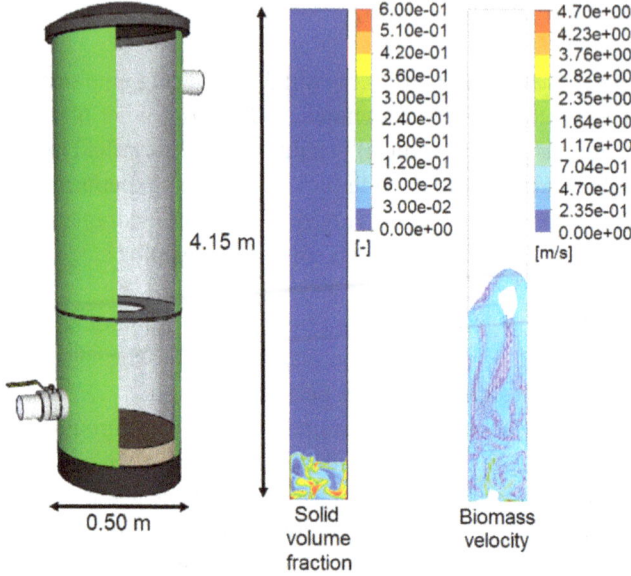

Figure 3.1: CFD representation of the solid volume fraction and biomass velocity vectors in a pilot-scale fluidized bed reactor.

$$\frac{\partial}{\partial t}\left(\alpha_g \rho_g\right) + \nabla \cdot \left(\alpha_g \rho_g \vec{v}_g\right) = S_{gs} \tag{3.30}$$

$$\frac{\partial}{\partial t}\left(\alpha_s \rho_s\right) + \nabla \cdot \left(\alpha_s \rho_s \vec{v}_s\right) = S_{sg} \tag{3.31}$$

where g and s indicate the gas and solid phases, respectively. The α is the phasic volume fraction, ρ is density, and $\vec{v}$ is velocity. When the solid phase is changed into the gas phase, the mass transfer, S, defined as the mass source term can be expressed according to eq. (3.32):

$$S_{sg} = -S_{gs} = M_c \sum Y_c R_c \tag{3.32}$$

To compute the gas phase density, the ideal gas behavior was considered (eq. (3.33)):

$$\frac{1}{\rho_g} = \frac{RT}{p} \sum_{i=1}^{n} \frac{Y_i}{M_i} \tag{3.33}$$

The solid phase density was assumed as constant.

3.2.1.2 Momentum equations

The momentum equations for gas and solid phases, respectively, can be written as

$$\frac{\partial}{\partial t}\left(\alpha_g \rho_g \vec{v}_g\right) + \nabla \cdot \left(\alpha_g \rho_g \vec{v}_g \vec{v}_g\right) = -\alpha_g \cdot \nabla p_g + \nabla \cdot \alpha_g \bar{\tau}_g + \alpha_g \rho_g \vec{g} + \beta\left(\vec{v}_g - \vec{v}_s\right) + S_{gs} U_s \quad (3.34)$$

$$\frac{\partial}{\partial t}\left(\alpha_s \rho_s \vec{v}_s\right) + \nabla \cdot \left(\alpha_s \rho_s \vec{v}_s \vec{v}_s\right) = -\alpha_s \cdot \nabla p_s + \nabla \cdot \alpha_s \bar{\tau}_s + \alpha_s \rho_s \vec{g} - \beta\left(\vec{v}_g - \vec{v}_s\right) + S_{sg} U_s \quad (3.35)$$

3.2.1.3 Turbulence model

The standard k–ε model in Ansys Fluent has become the workhorse of practical engineering flow calculations since it was proposed by Launder and Spalding [17]. It is a semiempirical model, and the derivation of the model equations relies on phenomenological considerations and empiricism. The selection of this turbulence model is appropriate when the turbulence transfer between phases plays a predominant role, as in the case of gasification in fluidized beds. The turbulence kinetic energy k and its rate of dissipation ε are obtained from the following transport equations:

$$\frac{\partial}{\partial t}(\rho k) + \frac{\partial}{\partial x_i}\cdot(\rho k u_i) = \frac{\partial}{\partial x_j}\left[\left(\mu + \frac{\mu_t}{\sigma_k}\right)\right] + G_k + G_b - \rho\varepsilon - Y_M + S_k \quad (3.36)$$

$$\frac{\partial}{\partial t}(\rho\varepsilon) + \frac{\partial}{\partial x_i}\cdot(\rho\varepsilon u_i) = \frac{\partial}{\partial x_j}\left[\left(\mu + \frac{\mu_t}{\sigma_\varepsilon}\right)\frac{\partial\varepsilon}{\partial x_j}\right] + C_{1\varepsilon}\frac{\varepsilon}{k}(G_k + G_{3\varepsilon}G_b) - C_{2\varepsilon}\rho\frac{\varepsilon^2}{k} + S_\varepsilon \quad (3.37)$$

3.2.1.4 Granular Eulerian model

The conservation equation for the granular temperature obtained from the kinetic theory of gases can be expressed as follows:

$$\frac{3}{2}\left[\left(\frac{\partial(\rho_s \alpha_s \Theta_s)}{\partial t}\right) + \nabla \cdot \left(\rho_s \alpha_s \vec{v}_s \Theta_s\right)\right] = \left(-P_s \bar{I} + \bar{\tau}_s\right):\nabla(\vec{v}_s) + \nabla(k_{\Theta_s}\nabla\Theta_s) - \gamma_{\Theta_s} + \varphi_{gs} \quad (3.38)$$

where γ_{Θ_s} is the collisional dissipation rate of granular energy given to the inelastic collisions, $\vec{v}_s$ is the diffusive flux of granular energy, φ_{gs} is the granular energy exchange between the gas and solid phases, and k_{Θ_s} is the diffusion coefficient.

By using eq. (3.39), given by Syamlal et al. [18], it is possible to compute the diffusion coefficient for granular energy:

$$k\theta_s = 15d_s/4(41-33\omega)\rho_s\alpha_s\sqrt{\theta_s\pi}\left[1+\frac{12}{5}\omega^2(4\omega-3)\alpha_s g_{0,ss} + \frac{16}{15\pi}(41-33\omega)\omega\alpha_s g_{0,ss}\right] \quad (3.39)$$

where $\omega = 1/2(1+e_{ss})$.

The granular energy dissipation can be computed by the following expression derived by Lun et al. [19]; here, solid pressure is considered for the pressure gradient term in the momentum equation, which includes a kinetic and a particle collision as given:

$$p_s = \alpha_s \rho_s \theta_s + 2\rho_s(1 + e_{ss})\alpha_s^2 g_{0,ss}\theta_s \tag{3.40}$$

3.2.1.5 Energy conservation model

$$\frac{\partial}{\partial t}\left(\alpha_q \rho_q h_q\right) + \nabla \cdot \left(\alpha_q \rho_q \vec{v}_q h_q\right) = \alpha_q \frac{\partial}{\partial t}(p_q) + \bar{\tau}_q : \nabla \cdot \vec{v}_q - \nabla \cdot \vec{q}_q + S_q$$

$$+ \sum_{p=1}^{n}\left(\dot{Q}_{pq} + \dot{m}_{pq}h_{pq} - \dot{m}_{pq}h_{pq}\right) \tag{3.41}$$

The energy conservation equation considers the heat exchange between gas and solid phases, viscous dissipation, and the void fraction's expansion work. To describe the energy conservation, the following energy conservation equation must be solved for each phase:

where $\dot{Q}_{pq}$ is the heat transfer intensity between gas and solid phases, h_q is the specific enthalpy of qth phase, $\vec{q}_q$ is the heat flux, S_q is the source term, and h_{pq} is the enthalpy of the interface. The heat transfer between the gas and solid phases can be written by considering the temperature difference between the gas and solid phases:

$$\dot{Q}_{pq} = h_{pq}\left(T_p - T_q\right) \tag{3.42}$$

The heat transfer coefficient relates to the particle Nusselt number of qth solid phase, and k_p is the thermal conductivity for pth phase, and is given by

$$h_{pq} = \frac{6k_p \alpha_q \alpha_p Nu_q}{d_p^2} \tag{3.43}$$

The Nusselt number is given by

$$Nu_s = \frac{h_{gs}d_s}{k_g} = \left(7 - 10\alpha_g + 5\alpha_g^2\right)\left(1 + 0.7Re_s^{0.2}Pr_g^{0.33}\right) + \left(1.33 - 2.4\alpha_g + 1.2\alpha_g^2\right)Re_s^{0.7}Pr_g^{0.33}$$

$$\tag{3.44}$$

where Re_s is the Reynolds number based on the diameter of the solid phase and the relative velocity, and Pr_g is the Prandtl number of the gas phase.

3.2.1.6 Chemical reactions model

Devolatilization is a process where moisture and volatile matters are driven out from biomass by heat. No devolatilization process is included in Ansys/Fluent considering the Eulerian–Eulerian method. To develop a reliable gasification model, it is necessary to include a devolatilization model in the Fluent code. Biomass is thermally decomposed into volatiles, char, and ash, in agreement with the following equation:

$$\text{biomass} \rightarrow \text{char} + \text{volatiles} + \text{steam} + \text{ash} \tag{3.45}$$

Note that ash appears for illustration only; the simulation does not consider ash content in the solid phase. In this work, the volatile matter is composed of the following species:

$$\text{volatiles} \rightarrow \alpha_1 CO + \alpha_2 CO_2 + \alpha_3 CH_4 + \alpha_4 H_2 \tag{3.46}$$

The biomass mixture composition is determined based on the proximate and elemental analyses shown in Table 3.3. Since there are no data on the exact distribution of the volatiles in biomass, the single-rate model developed by Badzioch and Hawsley [20] was adopted. The single-rate model produced moderate and reliable devolatilization rates with little computational effort. Additionally, a demoisturization equation was also considered. These two pseudo-heterogeneous reactions were modeled with a single reaction rate in agreement with the Arrhenius law.

3.2.1.7 Homogeneous gas-phase reactions

The modeling of the homogeneous gas-phase reactions should consider both the kinetic and the turbulent mixing rate effects [21]. The homogeneous reactions include only reactants and products in the same phase. In gasification, these reactions occur in the gas phase describing the reactions between volatile gases and gasifying agents. The gas phase is conditioned by the effect caused by the chaotic fluctuations of the solid particles. This turbulent flow leads to velocity and pressure fluctuations in the gaseous species. Fluent provides the finite-rate/eddy-dissipation model, which considers both the Arrhenius and the eddy-dissipation reaction rates. The water–gas shift reaction and the CO, H_2, and CH_4 combustion reactions were considered, respectively:

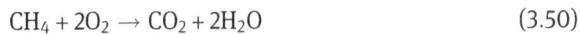

$$CO + H_2O \leftrightarrow CO_2 + H_2 \tag{3.47}$$

$$CO + 0.5O_2 \rightarrow CO_2 \tag{3.48}$$

$$H_2 + 0.5O_2 \rightarrow H_2O \tag{3.49}$$

$$CH_4 + 2O_2 \rightarrow CO_2 + 2H_2O \tag{3.50}$$

Table 3.3: Experimental operating conditions and syngas analysis.

Experimental conditions	Forest residues			Coffee husk			Vine-pruning residues		
Temperature (°C)	815	815	790	815	790	790	790	790	815
Admission biomass (kg/h)	63	74	63	28	28	41	25	55	55
Air flow rate (Nm³/h)	94	98	98	75	72	80	52	40	40
Syngas flow rate (Nm³/h)	106	94	100	106	88	116	107	108	102
Syngas fraction (dry basis)									
H₂	8.2	8.4	7.6	12.4	7.6	7.5	5.1	10.4	12.7
CO	18.6	18	17.9	11.4	11.1	10.6	8.3	11.7	14.1
CH₄	4.6	4.4	4.4	1.6	2.4	2.4	1.1	2.4	2.3
CO₂	16.7	17.1	17.1	18.7	17	18.5	16.5	20.1	17.9
N₂	48	48.2	49.2	52.3	54.2	55.2	56.4	51.2	49.1
Syngas NHV									
MJ/Nm³	5.16	5.02	4.93	3.34	3.2	3.07	1.99	3.46	4.02

The Arrhenius rates for each one of these reactions can be expressed as follows, respectively:

$$r_{\text{co-combustion}} = 1.0 \times 10^{15} \exp\left(\frac{-16,000}{T}\right) C_{CO} C_{O_2}^{0.5} \tag{3.51}$$

$$r_{H_2\,\text{combustion}} = 5.159 \times 10^{15} \exp\left(\frac{-3430}{T}\right) T^{-1.5} C_{O_2} C_{H_2}^{1.5} \tag{3.52}$$

$$r_{CH_4\,\text{combustion}} = 3.552 \times 10^{14} \exp\left(\frac{-15,700}{T}\right) T^{-1.5} C_{O_2} C_{CH_4} \tag{3.53}$$

$$r_{\text{water-gas shift}} = 2780 \exp\left(\frac{-1510}{T}\right) \left[C_{CO} C_{H_2O} - \frac{C_{O_2} C_{H_2}}{0.0265 \exp\left(\frac{3968}{T}\right)} \right] \tag{3.54}$$

The eddy-dissipation reaction rate can be expressed using the following equation:

$$r_{\text{eddy dissipation}} = \alpha_{i,f} M_{w,f} A \rho \frac{\varepsilon}{k} \min\left(\min_R \left(\frac{Y_R}{\alpha_{i,f} M_{w,R}}\right), B \frac{\sum_p Y_p}{\sum_i^N \alpha_{i,R} M_{w,f}}\right) \tag{3.55}$$

The minimum value of these two contributions can be defined as the net reaction rate.

3.2.1.7.1 Species transport equations

The local mass fraction of each specie Y is computed by using a convection–diffusion equation as follows:

$$\frac{\partial}{\partial t}(\rho Y_i) + \nabla(\rho Y_i \vartheta) = -\nabla \cdot J_i + R_i + S_i \tag{3.56}$$

where J_i is the diffusion flux of species i due to concentration gradients, R_i is the net generation rate of species i due to homogeneous reaction, and S_i is a source term related to the species i production from the solid heterogeneous reaction. The diffusion flux was computed as a function of the turbulent Schmidt number.

3.2.1.7.2 Heterogeneous gas-phase reactions

In the heterogeneous reactions, the gas species react with solid char (the solid devolatilization residue), resulting in gaseous products. When compared with the other gasification process steps (devolatilization and homogeneous reactions), the heterogeneous reactions have a much slower reaction rate. Thus, they are considered as the controlling step during the whole gasification process.

Heterogeneous processes occur at the interfaces where different phases meet. In these interface sites, a set of chemical reactions take place, known as reactions on surfaces. These reactions involve more than one phase, in which at least one of the steps of the reaction mechanism is the absorption of one or more reactants.

Char is the solid devolatilization residue. Heterogeneous reactions of char with the gas species such as O_2 and H_2O are complex processes that involve balancing the rate of mass diffusion of the oxidizing chemical species to the surface of biomass particles with the surface reaction of these species with the char. The oxygen diffusion determines the overall rate of a char particle to the particle surface and the rate of surface reaction, which depend on the temperature and composition of the gaseous environment and the size, porosity, and temperature of the particle. The commonly simplified reaction models consider the following overall reactions, such as char combustion (eq. (3.57)), H_2O (eq. (3.58)), and CO_2 char gasification (eq. (3.59)):

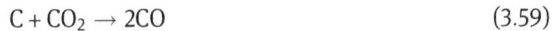

$$C + O_2 \rightarrow CO_2 \tag{3.57}$$

$$C + H_2O \rightarrow CO + H_2 \tag{3.58}$$

$$C + CO_2 \rightarrow 2CO \tag{3.59}$$

The heterogeneous reactions are influenced by many factors, namely, reactant diffusion, breaking up of char, interaction of reactions, and turbulence flow. To include both diffusion and kinetic effects, the kinetic/diffusion surface reaction model [22, 23] was applied. This model weights the Arrhenius rate's effect and the oxidant's diffusion rate at the surface particle. The diffusion rate coefficient can be defined as follows:

$$D_0 = C_1 \frac{\left((T_p + T_\infty)/2 \right)^{0.75}}{d_p} \tag{3.60}$$

where d_p represents particle diameter.

The Arrhenius rate can be defined as follows:

$$r_{\text{Arrhenius}} = Ae^{-\left(\frac{E}{RT_y}\right)} \tag{3.61}$$

All materials (gas species, solid biomass particles) were assigned appropriate properties from standard thermodynamic tables. The properties of the gas species (density ρ, viscosity μ, thermal conductivity k, and specific heat capacity C_p) were allowed to vary with local main phase temperature, and the mixture value was calculated from its local composition and available Fluent laws (ideal gas law for ρ and mass-weighted mixing law for μ, k, and C_p).

3.2.2 Expansion to municipal solid waste (MSW) residues

To deal with municipal solid waste (MSW) gasification, the devolatilization submodel must be upgraded considering the heterogeneous nature of MSW residues, mainly composed of plastics (polyethylene, polystyrene, and polypropylene, among others) and cellulosic materials (cellulose, hemicellulose, and lignin). Table 3.4 presents the chemical nature and composition of the MSW used in ANSYS Fluent to model the gasification process.

Table 3.4: MSW chemical nature and composition set to model the MSW gasification process.

Category	% Weight	Chemical formula
Cellulosic material	85.42	1
Polyethylene	10.99	$(C_2H_4)_n$
Polyethylene terephthalate	2.02	$(C_{10}H_8O)_n$
Polypropylene	0.81	$(C_3H_6)_n$
Polystyrene	0.76	$(C_8H_8)_n$

Pyrolysis reactions precede both homogeneous and heterogeneous reactions. Therefore, modeling pyrolysis is crucial for MSW gasification purposes. MSW is thermally decomposed into volatiles, char, and tar. There are several approaches to describe this phenomenon. Three main approaches are usually followed: a single-step pyrolysis model, competing parallel pyrolysis, and a pyrolysis model with a generation of secondary tar. Here, we adopt a pyrolysis model with the generation of secondary tar. The MSW is mainly composed of cellulosic and plastic components. The cellulosic material can be divided into cellulose, hemicellulose, and lignin [24, 25], and the plastics are mainly comprised of polyethylene, polystyrene, and polypropylene, among others. To distinguish the several components that comprise MSW, the pyrolysis reactions of cellulosic and plastic groups are considered individually and following an Arrhenius kinetic expression.

The primary pyrolysis equations can be defined as follows:

$$\text{Cellulose} \xrightarrow{r1} a_1 \text{volatiles} + a_2 \text{TAR} + a_3 \text{char} \tag{3.62}$$

$$\text{Hemicellulose} \xrightarrow{r2} a_4 \text{volatiles} + a_5 \text{TAR} + a_6 \text{char} \tag{3.63}$$

$$\text{Lignin} \xrightarrow{r3} a_7 \text{volatiles} + a_8 \text{TAR} + a_9 \text{char} \tag{3.64}$$

$$\text{Plastics} \xrightarrow{r4} a_{10} \text{volatiles} + a_{11} \text{TAR} + a_{12} \text{char} \tag{3.65}$$

The kinetics for the cellulosic material can be given as follows:

$$r_i = \frac{da_i}{dt} = A_i \exp\left(\frac{-E_i}{T_s}\right)(1 - a_i)^n \tag{3.66}$$

where i stands for cellulose, hemicellulose, and lignin (r_{1-3}); A_i is the pre-exponential factor, E_i is the activation energy; and n is the order reaction. The values for each one of these parameters can be found in [26]. Average values were considered.

Regarding the kinetic reactions for plastics, data were obtained from [27] by using the following reactions:

$$r_4 = \left[\sum_{i=1}^{n} A_i \exp\left(\frac{-E_i}{RT}\right)\right]\rho_v \tag{3.67}$$

where A_i, E_i, and ρ_v are the pre-exponential factor, the activation energy, and the volatile density, respectively, and can be found in [27]. i stands for each one of the plastics that comprise the analyzed MSW.

In this model, it is considered a secondary pyrolysis generating volatiles and secondary tar, as follows:

$$\text{Primary TAR} \xrightarrow{r5} \text{volatiles} + \text{Secondary TAR} \tag{3.68}$$

As this secondary pyrolysis is also very difficult to treat, a simplified global reaction is used [28]:

$$r_5 = 9.55 \times 10^4 \exp\left(\frac{-1.12 \times 10^4}{T_g}\right)\rho_{TAR1} \tag{3.69}$$

3.2.3 3D Eulerian–Eulerian approach

As seen, gasification systems encompass a set of phenomena such as fluid flow, heat and mass transfer, and complex chemical reactions. CFD models solve these process phenomena by employing a set of governing mathematical equations, mostly based on conservation equations, namely, mass, momentum, and energy. When employing a 2D or a 3D geometry domain, differences arise regarding their respective governing equations, given that the mathematical model considers the boundary threads of the applied domain. Therefore, only the differences between 2D and 3D momentum governing equations (in polar and rectangular form) are demonstrated for simplification reasons. As shown in 2D versus 3D momentum governing equations, x and y in the 2D expanded momentum equation refer to the r and y in the 3D cylindrical coordinate system and, naturally, to x and y in the 3D Cartesian system. The additional 3D terms are provided by θ (polar direction) in the 3D cylindrical coordinate system and z in the 3D Cartesian coordinate system. The expanded form of the vector gas momentum equation form in the 2D Cartesian system is given by [29]

$$\frac{\partial}{\partial t}\left(\alpha_g\rho_g v_g\right) + \frac{\partial}{\partial x}\left(\alpha_g\rho_g u_g v_g\right) + \frac{\partial}{\partial y}\left(\alpha_g\rho_g v_g v_g\right)$$

$$= -\frac{\partial P_g}{\partial y} + \frac{\partial}{\partial x}\left[\alpha_g\mu_g\left(\frac{\partial u_g}{\partial y} + \frac{\partial v_g}{\partial x}\right)\right] + \frac{\partial}{\partial y}\left(2\alpha_g\mu_g\frac{\partial v_g}{\partial y}\right) \tag{3.70}$$

$$-\frac{\partial}{\partial y}\left[\frac{2}{3}\alpha_g\mu_g\left(\frac{\partial u_g}{\partial x} + \frac{\partial v_g}{\partial y}\right)\right] + F_{gs}\left(v_s - v_g\right) - \alpha_g\rho_g g$$

where the subscript g represents the gas phase, v and u are the velocity components in the y-direction and x-direction, respectively. The terms $-\frac{\partial P_g}{\partial y}$ and $F_{gs}\left(v_s - v_g\right)$ relate to the pressure gradient and drag force.

The gas momentum equation form in the 3D cylindrical coordinate system goes as follows [29]:

$$\frac{\partial}{\partial t}\left(\alpha_g\rho_g v_g\right) + \frac{\partial}{\partial r}\left(\alpha_g\rho_g u_g v_g\right) + \frac{\alpha_g\rho_g u_g v_g}{r} + \frac{\partial}{\partial y}\left(\alpha_g\rho_g v_g v_g\right)$$

$$+ \frac{1}{r}\frac{\partial}{\partial \theta}\left(\alpha_g\rho_g v_g w_g\right)$$

$$= -\frac{\partial P_g}{\partial y} + \frac{\partial}{\partial r}\left[\alpha_g\mu_g\left(\frac{\partial u_g}{\partial y} + \frac{\partial v_g}{\partial r}\right)\right] + \frac{\alpha_g\mu_g}{r}\left(\frac{\partial u_g}{\partial y} + \frac{\partial v_g}{\partial r}\right) \tag{3.71}$$

$$+ \frac{\partial}{\partial y}\left(2\alpha_g\mu_g\frac{\partial v_g}{\partial y}\right) + \frac{1}{r}\frac{\partial}{\partial \theta}\left[\alpha_g\mu_g\left(\frac{1}{r}\frac{\partial v_g}{\partial \theta} + \frac{\partial w_g}{\partial y}\right)\right]$$

$$- \frac{\partial}{\partial y}\left[\frac{2}{3}\alpha_g\mu_g\left(\frac{\partial u_g}{\partial r} + \frac{\partial v_g}{\partial y} + \frac{1}{r}\frac{\partial w_g}{\partial \theta} + \frac{u_g}{r}\right)\right] + F_{gs}\left(v_s - v_g\right) - \alpha_g\rho_g g$$

where v, u, and w are the gas velocity components in the r-direction (radial), y-direction (streamwise) and θ-direction (azimuthal). Compared to eq. (3.70), two additional terms are shown on the left hand, $\frac{\alpha_g \rho_g u_g v_g}{r}$ and $\frac{1}{r}\frac{\partial}{\partial \theta}\left(\alpha_g \rho_g v_g w_g\right)$, and three additional terms on the right hand $\frac{\alpha_g \mu_g}{r}\left(\frac{\partial u_g}{\partial y} + \frac{\partial v_g}{\partial r}\right)$, $\frac{1}{r}\frac{\partial}{\partial \theta}\left[\alpha_g \mu_g\left(\frac{1}{r}\frac{\partial v_g}{\partial \theta} + \frac{\partial w_g}{\partial y}\right)\right]$, and $-\frac{\partial}{\partial y}\left[\frac{2}{3}\alpha_g \mu_g\left(\frac{\partial u_g}{\partial r} + \frac{\partial v_g}{\partial y} + \frac{1}{r}\frac{\partial w_g}{\partial \theta} + \frac{u_g}{r}\right)\right]$.

As for the 3D Cartesian coordinate system, eq. (3.70) appears as follows [29]:

$$
\frac{\partial}{\partial t}\left(\alpha_g \rho_g v_g\right) + \frac{\partial}{\partial x}\left(\alpha_g \rho_g u_g v_g\right) + \frac{\partial}{\partial y}\left(\alpha_g \rho_g v_g v_g\right) + \frac{\partial}{\partial z}\left(\alpha_g \rho_g v_g w_g\right)
$$

$$
= -\frac{\partial P_g}{\partial y} + \frac{\partial}{\partial x}\left[\alpha_g \mu_g\left(\frac{\partial u_g}{\partial y} + \frac{\partial v_g}{\partial x}\right)\right] + \frac{\partial}{\partial y}\left(2\alpha_g \mu_g \frac{\partial v_g}{\partial y}\right)
$$

$$
+ \frac{\partial}{\partial z}\left(\alpha_g \mu_g\left(\frac{\partial v_g}{\partial z} + \frac{\partial w_g}{\partial y}\right)\right)
$$

$$
- \frac{\partial}{\partial y}\left[\frac{2}{3}\alpha_g \mu_g\left(\frac{\partial u_g}{\partial x} + \frac{\partial v_g}{\partial y} + \frac{\partial w_g}{\partial z}\right)\right] + F_{gs}\left(v_s - v_g\right) - \alpha_g \rho_g g
$$

(3.72)

In comparison to the 2D Cartesian system, eq. (3.70), one additional term is placed on the left-hand side for the z-component, $\frac{\partial}{\partial z}\left(\alpha_g \rho_g v_g w_g\right)$, and two additional terms on the right-hand side, $\frac{\partial}{\partial z}\left(\alpha_g \mu_g\left(\frac{\partial v_g}{\partial z} + \frac{\partial w_g}{\partial y}\right)\right)$, and $\frac{\partial}{\partial y}\left[\frac{2}{3}\alpha_g \mu_g\left(\frac{\partial w_g}{\partial z}\right)\right]$. All other terms are common to all three eqs. (3.70), (3.71), and (3.72). Table 3.5 summarizes the 2D Cartesian and 3D cylindrical coordinate systems [30].

As an example, Figure 3.2 shows the 2D and 3D instantaneous quartz sand and eucalyptus wood biomass volume fraction contours, jointly with the 2D and 3D bed expansion at different superficial gas velocities (0.25, 0.40, and 0.60 m/s). Although both configurations deliver proper solid distribution, a sharper representation of solid separation along the bed height is provided by the 3D configuration, while the 2D considers a fade-out behavior between the solid separation. This may be given to the more accurate and realistic behavior analysis broadly addressed to 3D simulations [30]. The dimensionality effect is also quite visible in the bed expansion, with the 2D configuration tendentiously overestimating the bed expansion compared to the 3D configuration, particularly at higher velocities of 0.6 m/s (around 30% more). Such behavior is justifiable as the wall area, and the reactor volume ratio is larger for the 3D configuration; thus, the overestimation provided by the 2D configuration deems reasonable [30].

Besides these applications, the applied mathematical model was further extended to deal with gasification processes using CO_2 as a gasifying agent and multiple biomass substrates and to assess the hydrodynamics scale-up effect between two different-sized pilot-scale reactors (250 and 75 kW_{th}). Further details concerning these applications can be found elsewhere [16, 31, 32].

Table 3.5: Summary of the 2D Cartesian and 3D cylindrical governing equation terms.

2D Cartesian	3D cylindrical
	$\dfrac{\alpha_g \rho_g u_g v_g}{r}$
	$\dfrac{1}{r}\dfrac{\partial}{\partial \theta}\left(\alpha_g \rho_g v_g w_g\right)$
	$\dfrac{\alpha_g \mu_g}{r}\left(\dfrac{\partial u_g}{\partial y} + \dfrac{\partial v_g}{\partial r}\right)$
	$\dfrac{1}{r}\dfrac{\partial}{\partial \theta}\left[\alpha_g \mu_g\left(\dfrac{1}{r}\dfrac{\partial v_g}{\partial \theta} + \dfrac{\partial w_g}{\partial y}\right)\right]$
	$\dfrac{\partial}{\partial y}\left[\dfrac{2}{3}\alpha_g \mu_g\left(\dfrac{\partial u_g}{\partial r} + \dfrac{\partial v_g}{\partial y} + \dfrac{1}{r}\dfrac{\partial w_g}{\partial \theta} + \dfrac{u_g}{r}\right)\right]$
Common terms:	
$-\dfrac{\partial P_g}{\partial y}$	$-\dfrac{\partial P_g}{\partial y}$
$\dfrac{\partial}{\partial t}\left(\alpha_g \rho_g v_g\right)$	$\dfrac{\partial}{\partial t}\left(\alpha_g \rho_g v_g\right)$
$F_{gs}\left(v_s - v_g\right)$	$F_{gs}\left(v_s - v_g\right)$

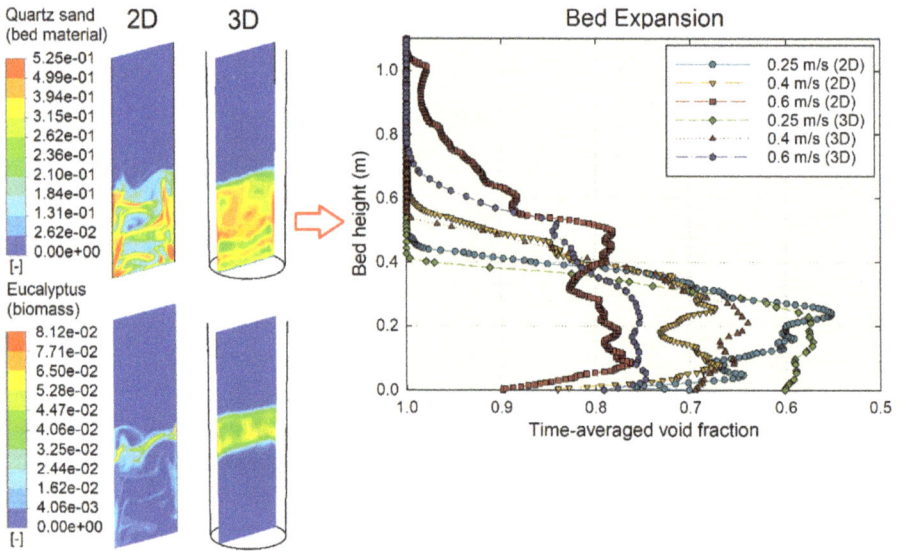

Figure 3.2: 2D and 3D instantaneous quartz sand and eucalyptus wood volume fractions.

3.2.4 Gasification optimization through CFD coupled with design of experiments (DoE)

Gasification is a highly complex process given the interplay of several involved phenomena such as hydrodynamics, mass transfer, momentum, energy balance, and chemical reactions [33]. Therefore, an experimental investigation is vital to understand the different entangled phenomena and operational parameters involved in the gasification process while assessing the optimal operating conditions and how these affect the gasifier's performance and syngas composition.

Given the intricate nature and the numerous factors participating in the gasification process, employing a conventional one-factor-at-a-time (OFAT) experiment approach requires both strenuous and time-consuming experimentation, neglecting multiple-factor interactions while being unfit to guarantee optimum method conditions [34]. Optimized operation conditions for complex processes such as gasification can be accomplished by employing advanced statistical approaches like design of experiments (DoE) coupled with the Monte Carlo method. DoE allows identifying the most robust combinations of factors by providing a cost-effective and time-efficient experimentation strategy, requiring far fewer experimental runs than traditional OFAT while simultaneously considering the effects of multiple factors on a given response instead of one at a time [35]. Monte Carlo leverages this optimization process, providing a more reliable and robust estimate by assessing the level of uncertainty [36]. In addition, when combined with DoE, this numerical method allows measuring each factor's variation impact on the system's overall performance while investigating the variation of all other involved parameters.

3.2.4.1 Experimental design

Several factorial designs are the most known: the 2k and 3k options and corresponding derivations. The 2k designs mean that there are two levels for each k factor and imply an approximately linear response over the range of the factor levels selected. This kind of design is quite usual in the industry and is particularly helpful in the first stage, where a large number of factors could be presented. Whenever, the developed models suggest a surface with curvature, the two-level designs will no longer give adequate information, and the 3k designs are the solution. At these circumstances, additional runs are required at new levels of input factors. The simplest approach goes through adding center points to a 2k design. If one notices significant variation, then it is wise to add new axial points beyond more center points. Sometimes, the range of selected levels is insufficient, and the user could assess outside the factorial levels further to assess the curvature. This last approach brings two possible drawbacks: going out from the factorial box could be physically impossible or even unsafe and hit too many levels and consequent too many runs [37, 38].

Any decision should consider the study specifications, the stage of experimentation, the factors restriction, and the possible model linearity or curvature. In general, typical experiments coming out from the industry lie under quantitative factors such as the temperature or the pressure. This means that these factors are adjusted to any level within the selected range. However, the reader could intend to perform a set of runs using different substrates or even catalysts. These factors are not quantitative but qualitative or categoric factors in the DoE jargon. With categoric factors, the user could select only one substrate or another but nothing in the middle. The combined use of qualitative and quantitative factors makes the process more complex, as the reader can guess.

The optimization procedure was engaged considering the results gathered from previous works developed by the author's research team for MSW gasification with air–CO_2 mixtures [31]. Overall, nine computer simulations were performed using the equivalence ratio (ER) and the CO_2–MSW ratio (CDMR) as input. Table 3.6 presents the full factorial design implemented with all possible selected factor combinations and responses. All remaining operating conditions were kept constant.

Table 3.6: Design for the MSW gasification.

	Gasification runs	ER	CDMR
MSW/air–CO_2	1	0.15	0.2
	2	0.25	0.2
	3	0.35	0.2
	4	0.15	0.5
	5	0.25	0.5
	6	0.35	0.5
	7	0.15	0.8
	8	0.25	0.8
	9	0.35	0.8

In this case, one is coupling the DoE with computational results from CFD simulations. As is easily understandable, CFD simulations always provide the same results, and the concept of experimental error does not make sense. Despite that, it is important to state some considerations about the experimental error when the reader can perform experimental runs. There is often a clear trend from people who are carrying out the experimental runs to avoid or eliminate the replication of center points. This could be a great mistake, and the reader is strongly advised not to do it. Excluding results coming from computational simulations, all the other data should provide for testing of lack of fit (LOF) [39]. LOF presents a relationship between the variation of the replicates and the variation of the design points about their predicted values. An F representation for the LOF test can be as follows:

$$F = \frac{\text{variation between the actual values and the values predicted from the model}}{\text{variation within any replicates}}$$

(3.73)

The lack of such a test compromises the awareness of how the model aligns with the current response data. To get such information, measuring pure error from replication is necessary. To avoid a large number of experimental runs, many companies only replicate the center points. Under these circumstances, they should get at least four replicates of the center point. However, random researchers through DoE papers reveal that still a significant number of researchers who do not follow such procedures lead to biased results.

When the user after computing the LOF test detects the model LOF, this could mean any one of two things: (a) replicates were run as repeated measurements leading to underestimated pure error, or (b) replicates have been run correctly, and the model does not fit the design points well.

Sometimes, due to time and laboratory restrictions, the center points are gathered in consecutive runs leading to lower errors than expected about the center point. When such occurs, the model fitting must be confirmed by applying a battery of additional statistical diagnosing tests. If such tests indicate that the model is robust enough to fit the empirical model under the navigation space, then the LOF test is no longer a significant test, and decisions about the model adequacy should be judged using another statistical analysis.

3.2.4.2 DoE single optimization

After selecting the most suitable experimental design, the reader is yet far from generating the response surface methodology (RSM) plots. First and foremost, there are some obligatory steps to take to ensure that the empirical model is highly predictable within the experimental space. The first step systematically determines how far it is worth going in the polynomial response (full quadratic or cubic polynomials are likely solutions). The sequential model sum of squares (SMSS) is a straightforward method to identify the high-level source of terms responsible for a significant variation in the response. More details about this method can be found in [40]. Table 3.7 turns out that the quadratic source includes all significant variations for the response. Using higher sources such as the cubic model is unnecessary, and the response is aliased. Note that there are additional statistical tests to confirm the right source level. Before showing these statistical measures, it is important to mention that the development of such an empirical model is based on Eulerian–Eulerian simulations under the CFD framework. Any computational-based simulation always provides the same solution for a set of input factors, flawing the concept of replicates. As mentioned earlier, using some statistical measures such as LOF does not bring any data of interest in such cases. However, measures such as R^2, R^2_{adj}, and R^2_{pred} are still useful. The R^2 measures how well the model can correctly fit the experimental data or the computed-based simulations as in the present case. The R^2 value can sometimes be misleading, causing overfitting of the data. The R^2_{adj} counteracts this overfitting giving a more reliable tool to evaluate the fitting data quality. The R^2_{pred} measures how well the model can refit the data when one point is missing. When these measures are close enough, a high-quality fit is expected. Table 3.8 reveals the R measures for the different sources, and the results align with the SMSS approach. Despite the cubic model presenting high R values, this is not a feasible option as it is aliased. The quadratic model stands out as the most valuable option with considerable high values for all R measures. Therefore, it is now wise to use a quadratic source for the response.

Table 3.7: Sequential model sum squares (CO_2 generation).

Source	Sum of squares	F-value	P-value prob < F
Quadratic vs 2FI	6.02	350.35	<0.0001
Cubic vs quadratic	0.071	4.81	0.0125

Table 3.8: Model summary statistics.

Source	R^2	R^2_{adj}	R^2_{pred}
Quadratic	0.9998	0.9998	0.9996
Cubic	1	0.9999	0.9997

The higher order polynomial determined by the SMSS suggested that the quadratic model provided the best fitting to the experimental data, including linear (A, B), interaction (AB), and quadratic (A^2, B^2) terms to depict the variation of the process. A and B terms suited the ER and CMDR factors, respectively, for the air–CO_2 mixture gasification. A final equation given by the quadratic model to predict the CO_2 generation for the MSW gasification process with air–CO_2 mixtures is given in eq. (3.74). Similar analysis and equalization were carried out for all remaining selected responses and gasification processes:

$$CO_2 = 24.12 + 1.33A + 3.35B + 0.25AB - 0.13A^2 + 0.32B^2 \tag{3.74}$$

This equation is in the coded form, and it is useful to identify the relative impact of factors by the signal and magnitude of the coefficients. Positive coefficients mean that increasing the factor leads to an increase in response and their magnitude aligns with their impact on the response. By default, the high levels of factors are coded as +1, and the low levels of factors are coded as –1.

The generation of the RSM plots still needs the last step, where the reader should diagnose the residuals or deviations from experimental points. Figure 3.3 shows the plot of experimental points for the selected model (CO_2 and other gases) versus the numerical calculation. Note that the plot lines up as expected, and the points hit the diagonal line. The reader has a large list of other diagnostic plots, but the previous steps followed in the above sequential order will ensure a great level of confidence in the picked model.

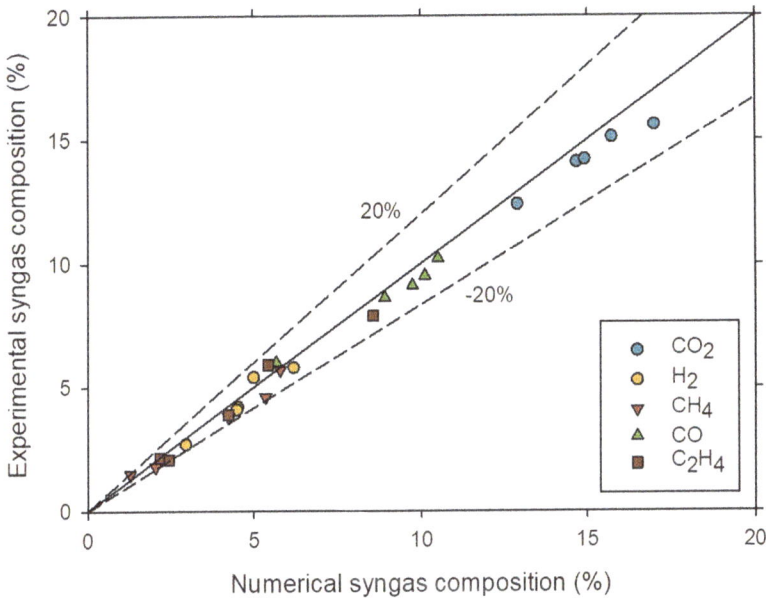

Figure 3.3: Comparison between experimental and numerical results.

The most straightforward way to evaluate the responses is to examine a contour plot of the fitted model. Figure 3.4 shows the contour plot for the single response. When there are only two or three factors under consideration, the interpretation of this kind of plot is easy. However, the process could become more complex when the user wants to explore a bunch of variables.

It is helpful first to evaluate the single responses without the interaction of the other parameters. This provides relevant information to select the best operating conditions to optimize each one of the responses. Table 3.9 reveals the operating conditions where this response hits the optimum to consider the applied design.

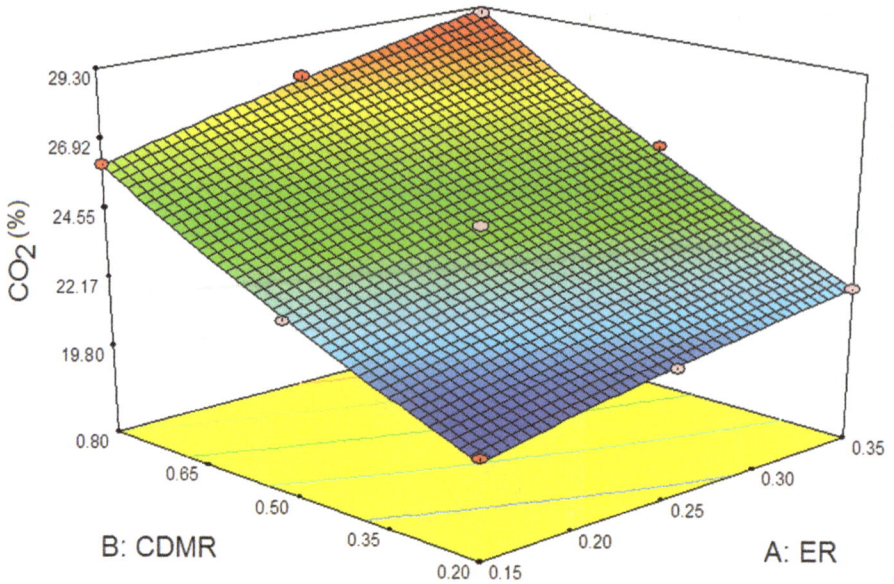

Figure 3.4: Contours for the CO_2 response.

Table 3.9: Maximum optimization prediction values for air gasification.

	Operating conditions for maximized response		
Response	CDMR	ER	Prediction
CO_2 (%)	100	75	29.28

3.2.4.3 Robust operating conditions

Having performed the single optimization, one reaches the desired operative target range necessary to obtain an optimized response. Still, some of these target points within the optimal range conditions may set the system on an acute response peak. Under such conditions, one may perform a robust optimization to deal with the uncertainty in the optimization data by selecting the set of operating conditions more robust to the variation imposed by the input factors. This is achieved by coupling the Monte Carlo method. A normal distribution requiring the input of a mean value and standard deviation was considered to accomplish the Monte Carlo analysis. This sort of probability distribution is often applied to DoE studies since it produces a realistic behavior of the considered uncertainties [41]. Within this normal distribution, the values were randomly generated under the distribution around the mean value up to their respective standard deviation allowing to compute the error created throughout the experimental area. This Monte Carlo simulation process is then repeated for a total of 5000 iterations, approximating the probability distribution of the final result. The employed standard deviations from each of the selected input factors needed to implement the method are given in Table 3.10. The data source considered was obtained from the experimental historical data gathered from the 250 kW_{th} pilot-scale gasification plant, as shown in [42].

Table 3.10: List of standard deviations for each considered input factor.

Input factors	Standard deviation
ER	0.01
CDMR	0.01

Figure 3.5 presents the 3D plot of the CO_2 generation (at optimal conditions) response error variance for air–CO_2 mixture gasification as a function of their respective input factors. Identical error variance trends were obtained for the remaining responses. Therefore, for simplification purposes, only the CO_2 responses for both processes are provided. This 3D plot combines the optimization procedure with robust conditions. It depicts the most stable operating conditions to achieve the maximum performance of the CO_2 response by providing the operative range at which the response error variance is diminished. The target region to achieve maximum performance is located at the mid-range of their input factors, while their minimum and particularly their maximum ranges return the highest error variances. Major evidence delivered by this kind of information is that the optimal operating conditions required to reach the maximum response earlier determined by single optimization may not always suggest the most stable set of values to operate the system. This is observable from Figure 3.5, as the maximum error variance meets with the optimal operating conditions to obtain a maximized CO_2 generation response, meaning that, in this study, a maximum response condition is not the most stable operating state to work with. This assumption reinforces the need to combine DoE with the Monte Carlo method.

Undeniably, this sort of information is particularly valuable in an industrial environment once it helps professionals understand the level of impact that each factor has on the overall performance of the system while simultaneously acknowledging the variation of all other involved parameters. In this case, combining DoE with the Monte Carlo method grants professionals working in gasification with a set of highly valuable tools and information, allowing them to make more secure, reliable, and smarter decisions considering the impact on a wide range of factors.

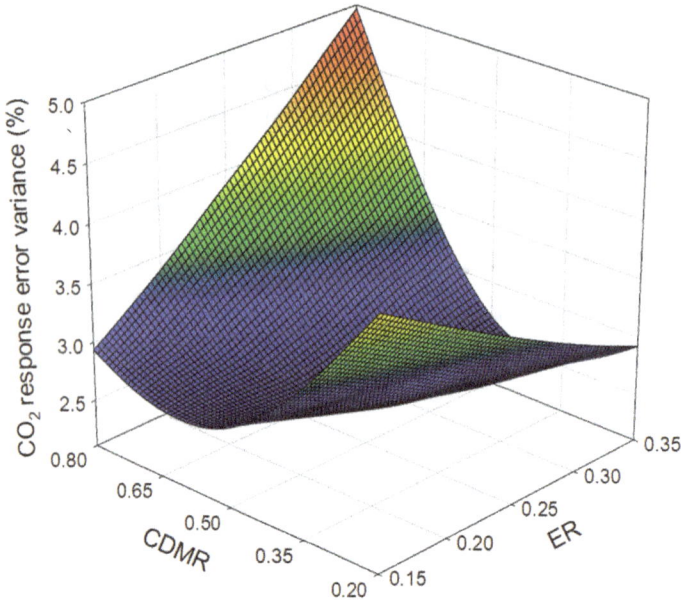

Figure 3.5: CO_2 response variance.

References

[1] V. B. Silva and A. Rouboa, "Using a two-stage equilibrium model to simulate oxygen air enriched gasification of pine biomass residues," *Fuel Processing Technology*, vol. 109, pp. 111–117, May. 2013, doi: 10.1016/J.FUPROC.2012.09.045.

[2] V. Silva and A. Rouboa, "Optimizing the gasification operating conditions of forest residues by coupling a two-stage equilibrium model with a response surface methodology," *Fuel Processing Technology*, vol. 122, pp. 163–169, Jun. 2014, doi: 10.1016/J.FUPROC.2014.01.038.

[3] V. B. Silva and A. Rouboa, "Predicting the syngas hydrogen composition by using a dual stage equilibrium model, " *International Journal of Hydrogen Energy*, vol. 39, no. 1, pp. 331–338, Jan. 2014, doi: 10.1016/J.IJHYDENE.2013.10.053.

[4] R. Desrosiers, "Thermodynamics of gas-char reactions," *A Survey of Biomass Gasification: Principles of Gasification*, vol. 2, 1979.

[5] J. M. Double and A. V. Bridgwater, "Sensitivity of theoretical gasifier performance to system parameters," 1985, pp. 915–919.

[6] M. J. Prins, K. J. Ptasinski, and F. J. J. G. Janssen, "Thermodynamics of gas-char reactions: First and second law analysis," *Chemical Engineering Science*, vol. 58, no. 3–6, pp. 1003–1011, 2003, doi: 10.1016/S0009-2509(02)00641-3.

[7] S. Jarungthammachote and A. Dutta, "Thermodynamic equilibrium model and second law analysis of a downdraft waste gasifier, " *Energy*, vol. 32, no. 9, pp. 1660–1669, Sep. 2007, doi: 10.1016/J.ENERGY.2007.01.010.

[8] R. T. Balmer, *Thermodynamics*, New York: West Publishing Company, 1990.

[9] Y. A. Cengel and M. A. Boles, *Thermodynamics: An Engineering Approach*, McGraw-Hill, 2002.

[10] M. V. N. H. Abbott, *Thermodynamics*, McGraw-Hill, 1972.

[11] S. A. Channiwala and P. P. Parikh, "A unified correlation for estimating HHV of solid, liquid and gaseous fuels, " *Fuel*, vol. 81, no. 8, pp. 1051–1063, May. 2002, doi: 10.1016/S0016-2361 (01)00131-4.

[12] R. Karamarkovic and V. Karamarkovic, "Energy and exergy analysis of biomass gasification at different temperatures, " *Energy*, vol. 35, no. 2, pp. 537–549, Feb. 2010, doi: 10.1016/J. ENERGY.2009.10.022.

[13] V. B. R. E. Silva and J. Cardoso, "Computational fluid dynamics applied to waste-to-energy processes a hands-on approach," 2020.

[14] N. Couto, et al., "Numerical and experimental analysis of municipal solid wastes gasification process," *Applied Thermal Engineering*, vol. 78, pp. 185–195, Mar. 2015, doi: 10.1016/J. APPLTHERMALENG.2014.12.036.

[15] J. Cardoso, V. Silva, D. Eusébio, P. Brito, M. J. Hall, and L. Tarelho, "Comparative scaling analysis of two different sized pilot-scale fluidized bed reactors operating with biomass substrates," *Energy*, vol. 151, pp. 520–535, May. 2018, doi: 10.1016/J.ENERGY.2018.03.090.

[16] J. Cardoso, V. Silva, D. Eusébio, P. Brito, M. J. Hall, and L. Tarelho, "Comparative scaling analysis of two different sized pilot-scale fluidized bed reactors operating with biomass substrates," *Energy*, vol. 151, pp. 520–535, May. 2018, doi: 10.1016/J.ENERGY.2018.03.090.

[17] B. E. Launder and D. B. Spalding, *Lectures in Mathematical Models of Turbulence*, vol. 176, Academic Press, 1972.

[18] M. Syamlal, W. Rogers, and T. J. O'brien, "(DE94000087) MFIX Documentation Theory Guide Technical Note," 1993.

[19] C. K. K. Lun, S. B. Savage, D. J. Jeffrey, and N. Chepurniy, "Kinetic theories for granular flow: Inelastic particles in Couette flow and slightly inelastic particles in a general flowfield," *Journal of Fluid Mechanics*, vol. 140, pp. 223–256, 1984, doi: 10.1017/S0022112084000586.

[20] S. Badzioch and P. G. W. Hawksley, "Kinetics of thermal decomposition of pulverized coal particles," *Industrial and Engineering Chemistry Process Design and Development*, vol. 9, no. 4, pp. 521–530, 1970, doi: 10.1021/I260036A005/TITLE/ KINETICS_OF_THERMAL_DECOMPOSITION_OF_PULVERIZED_COAL_PARTICLES.

[21] L. Yu, J. Lu, X. Zhang, and S. Zhang, "Numerical simulation of the bubbling fluidized bed coal gasification by the kinetic theory of granular flow (KTGF)," *Fuel*, vol. 86, no. 5–6, pp. 722–734, Mar. 2007, doi: 10.1016/J.FUEL.2006.09.008.

[22] M. M. Baum and P. J. Street, "Predicting the combustion behaviour of coal particles," vol. 3, no. 5, pp. 231–243, 2007, http://dx.doi.org/10.1080/00102207108952290, 10.1080/ 00102207108952290.

[23] M. A. Field, "Rate of combustion of size-graded fractions of char from a low-rank coal between 1 200°K and 2 000°K," *Combustion and Flame*, vol. 13, no. 3, pp. 237–252, 1969, doi: 10.1016/0010-2180(69)90002-9.

[24] R. C. Baliban, J. A. Elia, and C. A. Floudas, "Toward novel hybrid biomass, coal, and natural gas processes for satisfying current transportation fuel demands, 1: Process alternatives, gasification modeling, process simulation, and economic analysis, " *Industrial & Engineering Chemistry Research*, vol. 49, no. 16, pp. 7343–7370, Aug. 2010, doi: 10.1021/IE100063Y/ ASSET/IMAGES/IE-2010-00063Y_M077.GIF.

[25] O. Onel, A. M. Niziolek, F. M. F. Hasan, and C. A. Floudas, "Municipal solid waste to liquid transportation fuels – Part I: Mathematical modeling of a municipal solid waste gasifier," *Computers & Chemical Engineering*, vol. 71, pp. 636–647, Dec. 2014, doi: 10.1016/J. COMPCHEMENG.2014.03.008.

[26] P. Grammelis, P. Basinas, A. Malliopoulou, and G. Sakellaropoulos, "Pyrolysis kinetics and combustion characteristics of waste recovered fuels, " *Fuel*, vol. 88, no. 1, pp. 195–205, Jan. 2009, doi: 10.1016/J.FUEL.2008.02.002.

[27] C.-H. Wu, C.-Y. Chang, J.-L. Hor, S.-M. Shih, L.-W. Chen, and F.-W. Chang, "On the thermal treatment of plastic mixtures of MSW: Pyrolysis kinetics, " *Waste Management*, vol. 13, no. 3, pp. 221–235, Jan. 1993, doi: 10.1016/0956-053X(93)90046-Y.

[28] M. L. Boroson, J. B. Howard, J. P. Longwell, and W. A. Peters, "Product yields and kinetics from the vapor phase cracking of wood pyrolysis tars, " *AIChE Journal*, vol. 35, no. 1, pp. 120–128, Jan. 1989, doi: 10.1002/AIC.690350113.

[29] N. Xie, F. Battaglia, and S. Pannala, "Effects of using two- versus three-dimensional computational modeling of fluidized beds: Part II, budget analysis, " *Powder Technology*, vol. 182, no. 1, pp. 14–24, Feb. 2008, doi: 10.1016/J.POWTEC.2007.09.014.

[30] J. Cardoso, et al., "Comparative 2D and 3D analysis on the hydrodynamics behaviour during biomass gasification in a pilot-scale fluidized bed reactor," *Renewable Energy*, vol. 131, pp. 713–729, Feb. 2019, doi: 10.1016/J.RENENE.2018.07.080.

[31] N. Couto, V. Silva, and A. Rouboa, "Municipal solid waste gasification in semi-industrial conditions using air-CO2 mixtures," *Energy*, vol. 104, pp. 42–52, Jun. 2016, doi: 10.1016/J. ENERGY.2016.03.088.

[32] N. D. Couto, V. B. Silva, E. Monteiro, A. Rouboa, and P. Brito, "An experimental and numerical study on the Miscanthus gasification by using a pilot scale gasifier," *Renewable Energy*, vol. 109, pp. 248–261, Aug. 2017, doi: 10.1016/J.RENENE.2017.03.028.

[33] X. Ku, H. Jin, and J. Lin, "Comparison of gasification performances between raw and torrefied biomasses in an air-blown fluidized-bed gasifier," *Chemical Engineering Science*, vol. 168, pp. 235–249, Aug. 2017, doi: 10.1016/J.CES.2017.04.050.

[34] H. Yu, H. Yue, and P. Halling, "Comprehensive experimental design for chemical engineering processes: A two-layer iterative design approach," *Chemical Engineering Science*, vol. 189, pp. 135–153, Nov. 2018, doi: 10.1016/J.CES.2018.05.047.

[35] M. Douglas, *Design and Analysis of Experiments*, vol. 18, no. 2, John Wiley & Sons, Ltd, 2001, doi: 10.1002/QRE.458.

[36] J. P. C. Kleijnen, "Design and analysis of Monte Carlo experiments," *Handbook of Computational Statistics*, pp. 529–547, 2012, doi: 10.1007/978-3-642-21551-3_18.

[37] P. J. Whitcomb and M. J. Anderson, "RSM simplified: Optimizing processes using response surface methods for design of experiments," *RSM Simplified*, Nov. 2004, doi: 10.4324/9780367806071.

[38] V. Silva, et al., "Multi-stage optimization in a pilot scale gasification plant," *International Journal of Hydrogen Energy*, vol. 42, no. 37, pp. 23878–23890, Sep. 2017, doi: 10.1016/j. ijhydene.2017.04.261.

[39] M. J. Anderson and P. J. Whitcomb, "DOE simplified: Practical tools for effective experimentation," *Quality and Reliability Engineering International*, vol. 17, no. 4, Jul. 2001, doi: 10.1002/QRE.376.

[40] V. Silva, D. Eusébio, J. Cardoso, M. Zhiani, and S. Majidi, "Targeting optimized and robust operating conditions in a hydrogen-fed proton exchange membrane fuel cell," 2017, doi: 10.1016/j.enconman.2017.10.053.

[41] N. Noguer, D. Candusso, R. Kouta, F. Harel, W. Charon, and G. Coquery, "A framework for the probabilistic analysis of PEMFC performance based on multi-physical modelling, stochastic method, and design of numerical experiments, " *International Journal of Hydrogen Energy*, vol. 42, no. 1, pp. 459–477, Jan. 2017, doi: 10.1016/J.IJHYDENE.2016.11.074.

[42] J. Cardoso, V. Silva, and D. Eusébio, "Process optimization and robustness analysis of municipal solid waste gasification using air-carbon dioxide mixtures as gasifying agent, " *International Journal of Energy Research*, vol. 43, no. 9, pp. 4715–4728, Jul. 2019, doi: 10.1002/ER.4611.

4 Co-gasification and waste-to-energy conversion

4.1 Waste to energy

Waste-to-energy (WtE) or energy-from-waste (EfW) is the generation of power or heat from the primary treatment of waste or the conversion of waste into a fuel commodity, such as methane, methanol, ethanol, or synthetic fuels [1].

WtE technologies include any waste treatment process that generates EfW in power, heat, or transportation fuels [2]. The various WtE methods that may be employed for the valorization of solid waste fall into three categories: (1) thermal (direct combustion and incineration), (2) thermochemical (torrefaction, pyrolysis, plasma, and gasification), and (3) biochemical (composting, ethanol fermentation, and anaerobic digestion). Thermal technologies typically convert waste directly into heat energy, whereas thermochemical and biochemical technologies convert waste into secondary energy carriers such as syngas, torrefied pellets, biogas, bioethanol, and bio-oil, which can then be burned (in furnaces, steam turbines, gas turbines, or gas engines) to produce heat or electricity. Transforming solid wastes into secondary energy carriers enables a cleaner and more effective energy extraction method [3, 4]. Figure 4.1 depicts the technologies available for WtE.

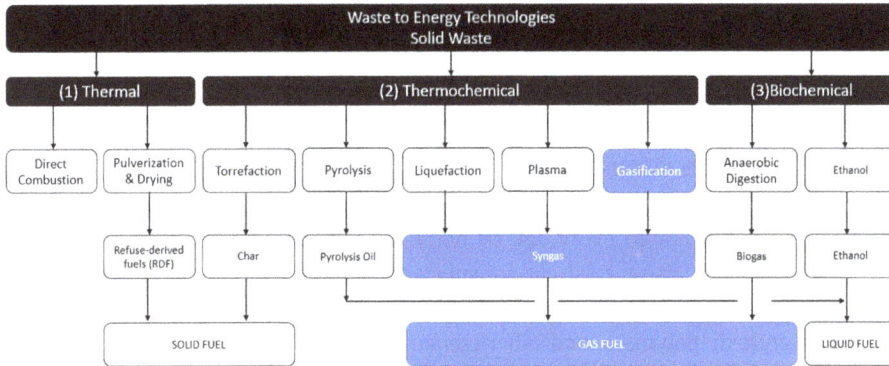

Figure 4.1: Waste-to-energy technologies.

Co-gasification of biomass and wastes has attracted the attention of several writers, demonstrating its appeal once a synthetic gas (syngas, also known as producer gas) with intriguing properties such as potential as a supplement or replacement for fossil-based ones is created [5, 6]. Moreover, co-gasification of mixes of biomass and plastic wastes, for example, has been proposed as a helpful technique for preventing issues that typically arise during the gasification of plastics alone, such as feeding difficulties and the production of pollutants [7, 8]. In addition, the drawbacks of either biomass or wastes may be mitigated, and the shortcomings of gasifying each

https://doi.org/10.1515/9783110758214-004

form of residue alone can be addressed [9–11]. Similarly, this kind of mixed fuels may assist in resolving issues associated with the unstable availability and variable composition of biomass, improving its exploitation and rationalization [12].

Unfortunately, competition for agroforestry residues is one of the biggest problems with gasification since agroforestry residues are also used for other things, like animal bedding, mushroom compost, strawberry production, straw mulching, and composite reinforcements [13, 14]. This situation may raise ecological and moral questions about how land is used, how it affects food production, and how natural resources are used to make it [13]. Also, more advanced methods that use agroforestry residues, such as making bioresins from sawdust lignin [15], offer competition. In addition, other ways are suggested and used to eliminate the use of municipal solid waste (MSW) and other types of waste, such as using waste plastic bottles in construction and making PET bottle bricks [16]. All these things can help cut down on MSW, but more needs to be done in this area.

To avoid environmental and moral concerns while getting the benefits listed earlier, blends of agroforestry residues and refuse-derived fuels (RDF) can be used. Using RDF is new and helpful since RDF is made up of MSW fractions and gives value to materials that would otherwise be thrown away in landfills, causing the problems we have already discussed and increasing the need for places to throw things away. So, using it can also reduce the amount of waste in sanitary landfills since these items would be sent to pyrolysis stations to be turned into fuel. In the same way, less plastic will end up in landfills, which is good because these materials are becoming a big problem because they take so long to break down naturally.

4.2 Solid waste conversion

MSW is the waste created every day by homes and businesses. It comprises mainly food waste, plastics, paper trash, and a substantial quantity of inorganic debris, such as metal scraps and glass [17]. In addition, a significant amount of MSW consists of biomass, synthetic polymers, and other combustible materials that may be utilized as an alternative energy source by implementing WtE systems, alleviating both waste and energy concerns.

Waste production is an inevitable outcome of wealth. As little towns grow into cities and cities grow into megacities, trash production rises. The greater availability of goods and services in urban areas encourages consumption tendencies and contributes to this transition toward higher trash production. Therefore, understanding responsible consumption is essential. In addition, various waste properties, activities, and strategies for collecting and recycling garbage come from diverse lifestyles [18, 19].

Forecasting global waste output is complicated because several aspects, such as country growth rates, population expansion, incomes, commercial agreements, new resources resulting from new technology and gadgets, conscious consumer awareness, and cultural considerations, must be considered. Figure 4.2 depicts the MSW generation share by world region and income level, with high-income countries having a gross domestic product (GDP) per capita greater than $12,476, upper middle-income countries having a GDP per capita between $4036 and $12,476, lower middle-income countries having a GDP per capita between $1026 and $4035, and low-income countries having a GDP per capita less than $1025 [20]. In addition, according to projections, worldwide trash output will reach 2.59 billion tons in 2030 and 3.50 billion tons in 2050 [19].

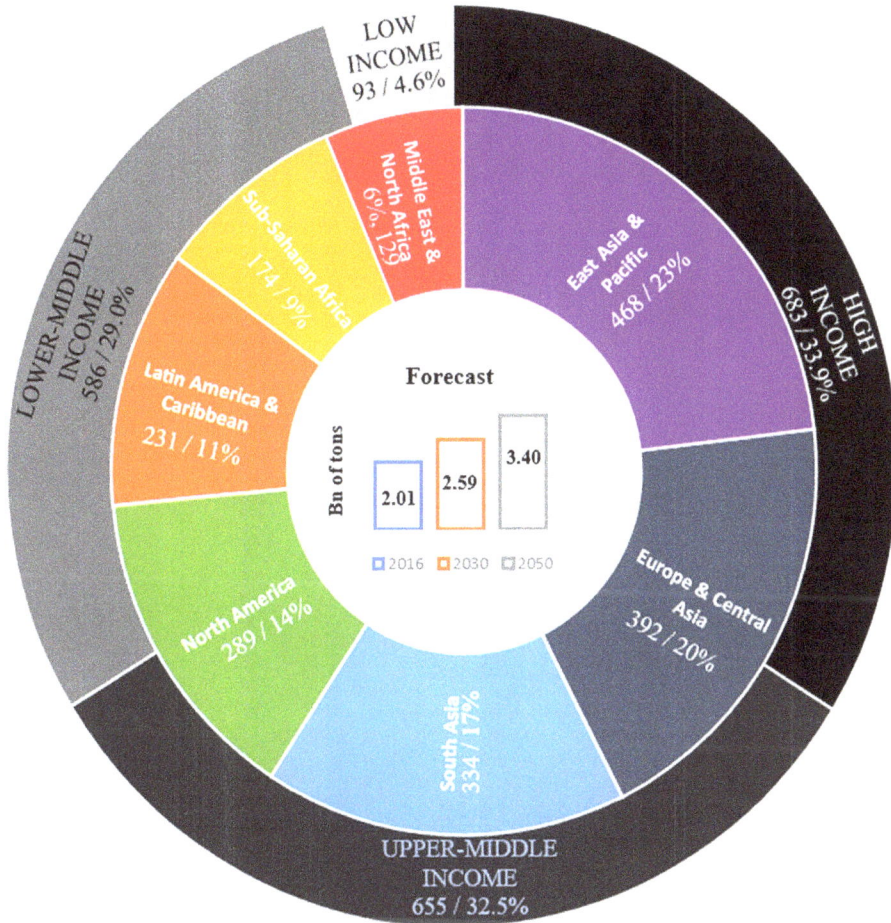

Figure 4.2: Waste production by income level and area in 2016 (millions of tons/percent) [9]. The outside circle indicates waste production by income level, the middle circle by region, and the inner circle is a projection of trash creation [19].

The income level of a society affects how much MSW is produced and what it is made of [19]. So, MSW is different in each place. This dramatically affects how MSW is treated and how much it can be reused and recycled. Figure 4.3 shows how waste is made up by the income level. In the high-income group, less food and green waste is made than in other groups. In high-income countries, on the other hand, there is more paper and cardboard, which could be a good thing because less pretreatment is needed to make RDF from MSW. But this pretreatment involves getting rid of moisture by drying. So, it takes more energy to eliminate MSW, which has a lot of food and green waste.

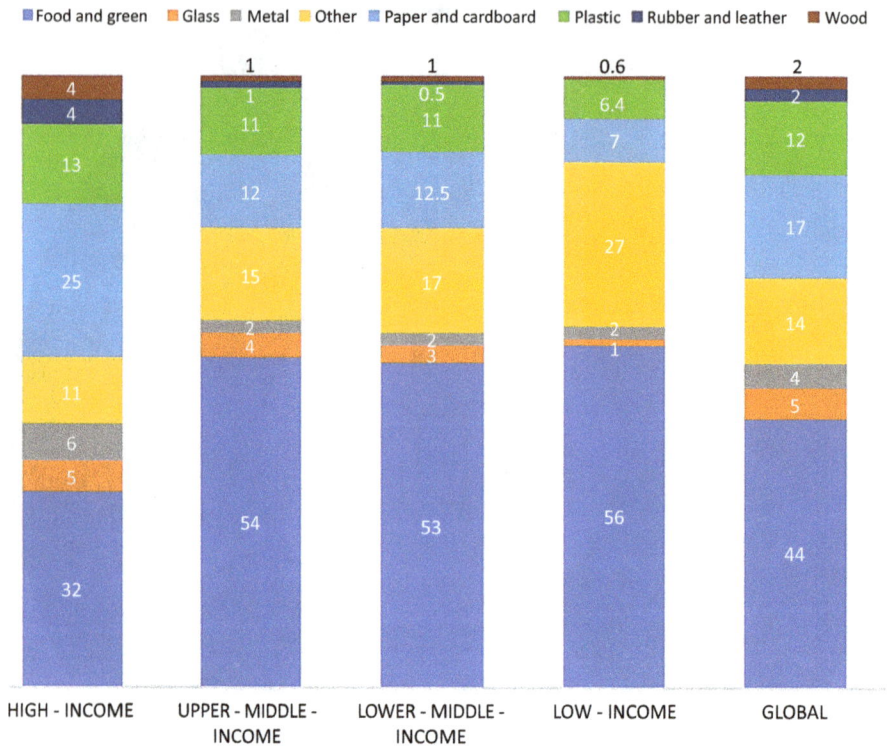

Figure 4.3: MSW composition based on income level.

MSW from high-income countries like the United States and Portugal has the properties needed to make RDF with less water than MSW from low-income countries. Some authors say that the high moisture content of MSW in developing countries makes it hard to make RDF, so they suggest a different biological treatment for the organic fraction [21]. The most significant difference in waste composition is between countries with high and low incomes, with different products and services available. This variety in MSW's makeup is essential when figuring out the best and most efficient way to produce RDF [22]. RDF can be used to make liquid and gaseous fuels by pyrolysis and gasification, respectively [23]. RDF can also be used as fuel in combustion.

New rules, standards, and environmental restrictions have confronted the solid waste management industry with significant hurdles. As a result, the conventional struggle over garbage incineration is experiencing a sea shift. Emerging subcategories such as innovative sanitation, waste sorting, and sophisticated technologies such as gasification and pyrolysis support strong business development [24]. The services provided by the leading solid waste management firms are detailed in Table 4.1. The column for the leading player includes the respective company's name, nation, and foundation year. In contrast, the business information column offers the ticker symbol and 52-week range price for listed firms (if applicable) (in 2020).

Some leading players have had spectacular development, such as Yingfeng Environmental Technology, whose stock price at the beginning of 2020 was about ¥6.5. By the end of July, it had increased to ¥10. China is the world's largest waste generator and is classified as a lower middle-income country. Thus, it relies heavily on open dumps and landfills for waste disposal. However, China is investing heavily in the WtE sector, and emerging companies are capitalizing on this opportunity by participating in and developing government-funded WtE projects [28].

4.3 Waste-to-energy thermochemical technologies

RDF can be used as a feedstock for thermochemical conversion processes like pyrolysis, gasification, and combustion that turn waste into useful energy sources [50]. For instance, combustion creates thermal energy (heat), which can be used in many ways, such as in industrial processes, district heating, and steam turbines that make electricity. Pyrolysis produces solids, liquids, and gases that can be turned into more valuable products. Finally, gasification makes a mixture of gases that can be burned. These gases can be used as fuel in a combined cycle to make electricity, or they can be used in advanced applications like Fischer-Tropsch to make chemicals or fuels.

4.3.1 Combustion

In combustion, the fuel is burned with an excess of oxygen at temperatures between 800 and 1000 °C, producing thermal energy in the form of a hot gas that can be used as process heat in industrial processes or converted to mechanical power in a steam turbine and then to electricity in a generator [29]. It is already a well-established and widely available commercial technology on domestic, small industrial, and utility scales [30]. This technology controls a sizable portion of the market.

Table 4.1: Top RDF and SRF, and MSW management players [25–27].

Top players	Company information	Services					
	Ticker symbol and 52-week range price	Waste collection	Disposal andrecycling	WtE	Hazardous treatment	Urban cleaning	RDF export 2018 (ton)
Capital Environment Holdings CN, 2004	3989:HK HK$0.10–0.24		X				
Jinjiang environment CN, NI	Private company			X			X
Yingfeng Environmental Technology CN, 1993	000967:CH ¥5.43–10.47	X	X	X	X	X	
TPI Polene Power TH, 1991	TPIPP:TB3 THB 2.76–6.30			X		X	
VeoliaFR, 1995	VIE: FP €16.02–19.28	X	X	X	X	X	158,206
PAPRECFR, 2006	Private company		X			X	
Advanced Disposal Services US, 2012	ADSW: US $25.88–33.15	X	X				
Clean Harbors, Inc. US, 1987	CLH: US $29.45–88.40	X	X		X		
Covanta Holding Corp US, 1992	CVA: US $7.00–18.38		X	X	X		X
Waste Management Inc. US, 1995	WM: US $85.34–126.79		X	X			X
ABRELPEBZ 1976	Private company	X	X	X	X	X	
3 R Management IN, –	Private company	X	X	X	X	X	
Biffa Group UK, 2000	BIFF: LN £165.40–314.00	X	X	X		X	350,167
SUEZ UK UK, 1978	Private company	X	X	X	X	X	302,393
Andusia Holdings UK, NI	Private company		X	X			146,620

Company	Ownership / financials						Capacity
Seneca Environmental Solutions UK, NI	Private company			X			302,393
Renewi PLC UK, 1982	RWI: LN £20.05–45.90			X			138,355
BMH Technology Oy FI, 2001	Private company					X	
Herambiente IT, –	Private company	X	X			X	
JFE Engineering JP, –	Private company	X	X	X		X	
Hitachi Zosen JP, 1934	Private company	X		X			
Beauparc Group IE, 1990	Private company	X	X	X		X	
Enva IE, NI	Private company	X	X	X	X	X	
N + P Group NL, 1992	Private company	X	X			X	334,154
FCC Magyarország Ltd HU, 1991	Private company	X		X			162,870

Table 4.2 summarizes several studies on combustion using RDF, MSW, and other residues as fuel, either alone or with other fuels (e.g., biomass or coal). The fluidized bed technology is primarily used here because these solid wastes and their derivatives ignite at relatively low temperatures and are highly reactive, implying that they can be used successfully as a fuel in these systems [31]. Additionally, Table 4.2 summarizes various studies demonstrating a reduction in ignition and volatile matter combustion temperature when samples are co-fired with a conventional feedstock (e.g., coal) and RDF. Combining conventional feedstocks with RDF reduces the activation energy required to burn the char. This effect reduces energy consumption and combustion time, making it more economically viable [31].

Table 4.2: Studies on RDF combustion and co-combustion.

Year	Feedstock	Equipment	Mixture (%)	Calorific values of feedstock (MJ/kg)	Activation energies (kJ/mol)
2020	Low-quality coal/RDF/ plastic-paper [31]	Fluidized bed combustor	Several	$21.7–16.7/31.2–22.4^{db}$	53–290
2019	SRF + mineral additive [32]	Fluidized bed combustor	100	$21.3–26.6^{wb}$ GCV	–
2018	MSW/coal [33]	Drop tube furnace	Several	$19.32/25.85^{db}$ NHV	–
2016	RDF A-B/coal/petroleum coke [34]	Fluidized bed combustor	Several	$22.4–19.22/25.74/4.38^{db}$ LHV	–
2013	RDF/sawdust [35]	Vortexing fluidized bed combustor	50/50	$15.78/14.03^{db}$ LHV	–
2011	SRF/coal/PVC [36]	Entrained flow reactor	Several	$20.8/26.5/19.0^{wb}$ LHV	–

MSW incineration is an excellent option to reduce the volume of waste in landfills and produce heat and electricity. However, WtE plants' electrical efficiency is set between 15% and 25% [37] due to the low temperature and pressure of the process used to minimize scaling and corrosion. The corrosion is produced by chloride (Cl^-) in MSW since it promotes the formation of corrosive compounds and volatile compounds formed with metal traces like zinc (Zn) and lead (Pb) [37]. Similar RDF contains chlorine in their composition because of plastics like PVC and biomass fractions. In this context, several measures can be taken to reduce chlorine compound emissions, including adding calcium compounds to dolomite and limestone and aluminosilicates to kaolin and halloysite [32].

RDF has a greater LHV than MSW. However, the manufacture of RDF requires energy for sorting, processing, and moisture reduction in some cases. This procedure will incur additional costs, which some writers believe are unaffordable [38].

4.3.2 Pyrolysis

Pyrolysis is defined as the thermochemical decomposition of carbonaceous fuel (e.g., biomass, MSW, and RDF) without air/O_2, as shown in the following equations:

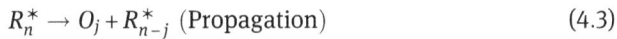

$$Biomass \rightarrow Char + Ash + Moisture + Volatile \ (CO, \ CO_2, \ CH_4, \ C_2H_4, \ H_2O) \quad (4.1)$$

$$Biomass_{Molecule} \rightarrow 2R^* \ (Initiation) \quad (4.2)$$

$$R_n^* \rightarrow O_j + R_{n-j}^* \ (Propagation) \quad (4.3)$$

$$2R^* \rightarrow Products \ (Termination) \quad (4.4)$$

The general pyrolysis reaction is the one shown in eq. (4.1). The process of thermal cracking is shown by the remaining reactions, where R_n^* is an n-chain free radical. For example, j is an alkene with a long chain from the olefins in RDF [39].

The pyrolysis starts at around 350 °C and goes up to 700 °C or higher, generating three main products: oil, gases, and solids (char). Pyrolysis has been used for a long time to produce charcoal from biomass, but in recent years has emerged with significant potential to generate oil from a set of biomass and solid wastes and their mixtures [40]. However, several factors affect how well pyrolysis works, such as the chemical makeup of the feedstock, the cracking temperature, the rate of heating, the type of reactor, the residence time, the use of catalysts, and the pressure [39]. Because of this, large-scale facilities need to consider how these factors affect the amount of product made and how well the process works.

Table 4.3 summarizes some RDF and other solid waste pyrolysis studies. Unfortunately, not all of them provide helpful information about the distribution and quality of their products such as char, gas, and bio-oils (e.g., chemical composition and heating value).

RDF pyrolysis and co-pyrolysis have been primarily studied in a fixed-bed reactor. In co-pyrolysis operation, the mass ratio between co-reactants (e.g., MSW and Olive Kernel) has an essential effect on the process, and the degree of contact between co-reactants influences the synergistic effect, which is more favorable for co-pyrolysis in a fixed bed reactor than in a fluidized bed reactor [41]. This effect may be due to the synergistic catalytic cracking effect of the inorganics present in the feedstock, often named ash [41].

As previously analyzed, MSW can be processed to produce RDF by removing a considerable number of materials like glass, metals, and biodegradable fractions, thus decreasing the inorganics and moisture content of the feedstock known to

Table 4.3: Studies on RDF pyrolysis and co-pyrolysis.

Year	Feedstock	Reactor bed	High heating value of feedstock (MJ/kg)	Mixture	Temperature (°C) and heating rate (°C/min)	Char/oil heating value (MJ/kg)	Gas heating value (MJ/Nm³)
2019	Olive mill wastewater sludge/ waste tires [41]	Fixed	24.64/40.60^{db}	Several	600 @ 20 °C/min	−/39.5–43 HHV	−
2018	Olive kernel/ MSW [42]	Fixed	20.6/15.6^{db}	Several	600–800	−/−	14.5–16.8 HHV
2017	Plastic waste [43]	Fixed	−	Several	450	−/41.4–41.8 HVV	−
2015	RDF [44]	Fixed	17.2^{db}	Several	500 °C @ 10 °C/min	−/32.9–41.3 HHV	17.1–21.9 HHV
2014	Waste tires/ forestry waste [45]	Fixed	38.6/19.5^{wb}	Several	500 @ 80 °C/min	−/−	11.4–59.0 LHV
2013	RDF [46]	Fixed	17.9^{db}	Several	500–900	−/−	9.4–10.4 LHV

have a detrimental effect on the pyrolysis process. Therefore, the process from MSW to RDF enhances the performance of pyrolysis toward higher liquid yields, as less moisture content means improved energy content of the pyrolytic liquids [47]. Furthermore, MSW is a heterogeneous material, leading to product quality variations. In contrast, RDF comprises a selective MSW fraction, which affords constant density and size, consistent composition, higher heating value (HHV), and transport ease [44].

Still, the pyrolysis of RDF has a recognized drawback of producing significant amounts of waxes, often related to the pyrolysis of plastic fraction, which can create problems during the process [48]. Therefore, some catalysts have been tested to reduce wax yields and improve bio-oil properties. For example, the ZSM-5 catalyst effectively reduces the number of waxes, producing higher deoxygenation and organic liquid yields [44]. Other low-cost catalysts include the char from RDF pyrolysis and oyster shell, which effectively reduce waxes and improve the deoxygenation of liquid pyrolysis products [44].

Chavando et al. [49] analyzed the pyrolysis of blends of RDF–wood and HDPE (high-density polyethylene)–wood. They highlighted three main aspects: (1) how the cracking temperature affects product yields; (2) how the heating rate affects product yields; and (3) how increasing the weight percentage of RDF and HDPE affects product yields. This information is essential for figuring out how to use waste materials (like RDF and HDPE) in the traditional pyrolysis of lignocellulosic materials like wood, which is an excellent way to deal with waste and make energy vectors. But pyrolysis of 100% RDF and HDPE makes a heavy molecular tar fraction (waxes) that condenses and hardens in the condensate system, causing blockages and other problems. So, mixing RDF and wood, as well as HDPE and wood, makes this problem much less severe.

Concerning how the cracking temperature affects the amount of product made, it was found that the amount of liquid made from RDF-wood and HDPE-wood blends is more significant at 500 °C. On the other hand, secondary cracking reactions took place at 550 °C, which made the gas yield go up. These reactions also reduce the amount of wax that plastics make, which is suitable for the setup because wax can clog the condensate system. The ashes also cut down the liquid yield in RDF. The rate of heating was the second effect that was looked at. It was noticed that as the heating rate goes up, the amount of liquids produced goes up, and the amount of char produced goes down, and vice versa. In the case of RDF blends, heating at a slower rate makes more char and less liquid. There was no significant change in the amount of liquid or char that HDPE blends made. But gas was increased at higher heating rates, which reduced the number of waxes suitable for HDPE blends. Adding HDPE and RDF improved the LHV of the gas, which was the last effect. For example, the LHV of gas made with only wood at 550 °C and 20 °C/min was 10.34 MJ/m^3, and it was 14.82 MJ/m^3 with 25 wt.% HDPE. These results show the potential and benefits of adding RDF and HDPE to industrial pyrolysis processes to make energy. As these materials are added, the LHV goes up, producing more energy. Using trash to make energy vectors is also a way to clean up the environment since trash has become a big problem.

4.3.3 Co-gasification with refuse-derived fuels (RDF)

By combining a solid with a gasifying agent (air, oxygen, carbon dioxide, steam, or their combinations) at high temperatures, often between 800 and 900 °C, gasification converts many solid carbonaceous feedstocks, such as coal, biomass, and solid waste, into a combustible gas mixture [50]. The work on this subject is summarized in Table 4.4, which highlights the gas quality based on the lower heating value (LHV) of gas. The LHV of various feedstocks is <20 MJ/kg, except for plastic and RDF combinations. The LHV of the generated gas is generally below 13 MJ/Nm^3. When air is utilized as the gasifying agent, the heating value is below 6 MJ/Nm^3, indicating that this is a low-quality gas (air usage leads to a gas highly diluted in N_2).

Blending RDF with other feedstocks improves the efficiency of the gasification process by increasing the concentration of CH_4 and C_2H_4 while decreasing the concentration of CO in the produced gas. This may be due to the combined effect of the

Table 4.4: Studies on RDF gasification.

Year	Feedstock	Reactor bed	Calorific values of feedstock (MJ/kg)	Bed material	Gasifying agent	Temperature (°C)	Lower heating value of gas(MJ/Nm^3)
2020	RDF/wood chips/wood pellets [51]	Fluidized	24.8/18.8/18^{db} LHV	Sand	Air	800	5.8–6.4
2020	RDF/wood [52]	Fluidized	17.1–17.5^{db} LHV	Sand	Air	800–850	4.6–5.4
2019	RDF [53]	Tubular	21.3^{wb} LHV	Mg catalyst	Air	700–850	–
2018	Pelletized MSW/chopped switchgrass [54]	Patented design	19.19/16.49 ^{b} LHV	–	Air	700–950	6.78–7.74
2017	RDF [55]	–	29.41^{db} HHV	–	Air	–	5.87
2016	PET pellets/wood pellets [56]	Fluidized	19.96/18.41^{wb} GHV	Sand	Air	725–875	3.4–5.1
2015	Sewage sludge/woody biomass [57]	Fixed	14.4/17.0^{db} HHV	–	Air	550–950	4.5
2014	HDPE/palm kernel shell [58]	Fluidized	45.98/24.97^{wb} HHV	Dolomite	Steam	650–800	–
2014	Plastic waste/lignite [59]	Fluidized	42.2/30.2^{db} HHV	–	Steam, CO$_2$	900	10–11.11
2013	Tire/almond shell/palm empty fruit [60]	–	19.05/14.13/17.09^{db} HHV	Natural catalysts	CO$_2$	850–1000	–
2012	Sewage sludge/forestry waste [61]	Fixed	–	–	Steam	700–900	11.89–12.72
2012	Sewage sludge/wood pellets [62]	Fixed	15.0/18.3^{db}	–	Air	550–850	3.5–6.0
2012	Plastic pellets/wood pellets/olive husk pellets [63]	Fluidized	21.9/18.5/19.3^{db} LHV	Quartzite sand/ 5.5% Ni	Air, steam	780	–
2011	Polyethylene/wood chips [64]	Fluidized	42.97/22.3 LHV	–	Air, steam	900	–

thermal cracking of the plastic polymers in the RDF and the catalytic effect induced by the feedstock's ashes [51]. Furthermore, blending RDF with biomass dilutes the RDF char (e.g., high ash and chlorine contents), allowing its energetic valorization in existing gasification facilities [52]. Furthermore, utilizing RDF in thermochemical conversion results in a homogeneous composition of the products, the produced gas.

The gas generated by gasification may be utilized for further applications, including the synthesis of methanol [65], ammonia production [66], biomethane [67], and liquid hydrocarbons [68].

Because RDF has a higher density, lower moisture content, and HHV than MSW, it is possible to avoid using large-scale equipment to process these feedstocks by substituting it for raw MSW. This has a positive economic impact because smaller equipment is less expensive to acquire and requires less maintenance. Moreover, this continuous scale-up process would open the path for decreased system manufacturing costs, enabling the broad deployment of large-scale systems, notably for pyrolysis and gasification. Table 4.5 details RDF plants, including implementation costs, longevity, input flow, energy production, and payback period.

The use of RDF in gasification affects the composition of the gas, efficiency, and equipment conditions. These effects are described further.

Bed material may also interact unfavorably with the fuel, altering its physical characteristics and leading to buildup, particularly in the case of biomass, whose combustion releases alkaline chemicals that, when in contact with silica bed material, create alkali silicates [72]. In this instance, natural rock bed materials (dolomite, olivine, limestone, etc.) might be substituted for silica since they are inexpensive and readily available, despite their lower mechanical strength, which can induce attrition [73–75]. In addition, alternate synthetic bed materials (such as alumina) may be used in place of the originals but at a higher cost. In situations where this replacement is not feasible owing to mechanical resistance, in-bed additives such as kaolin, calcium oxide or carbonite, and bauxite may aid in reducing buildup [76, 77]. In-bed additives are also a good alternative for tar reduction. For example, de Andrés et al. [78] observed that, given the applied circumstances, dolomite was the most active catalyst for tar removal from sewage sludge gasification. At the same time, olivine produced the least significant results.

Despite RDF having a higher density and lower moisture content than conventional feedstocks, they also have a considerable number of ashes or alkali and alkaline earth metals, which have a catalytic action on pyrolysis and carbon gasification [79, 80], with sodium and potassium playing key roles [81–83], among these metals. McKendry [84] reported the alkali metal content of a variety of biomass materials (including willow, cereal straw, bagasse, and switchgrass), and it can be concluded that this parameter (in terms of Na and K oxides) accounts for approximately 4–16% of the total composition of the selected biomass types. This alkali concentration provides biomass char with a greater surface area and porosity, increasing its gasification reactivity [85] and catalytic activity when combined with other wastes [60, 86].

Table 4.5: Example of combustion and gasification plants using RDF.

Technology	Country	Investment	Operating active period	Feedstock processing capacity	Energy output	Payback period (years)	Remarks
Combustion [69]	The Netherlands	€3,758,650	1986–1991	6.35 tons/h	29.620 MWe	7	–
Gasification [70]	Germany Operating	€7,563,452	2008–2015	500 kg/h45,000 tons/year	0.5 MWe	–	Gas: 5 MJ/Nm³ Secondary fuel (marketed under the name Stabilat)
Gasification [71]	Finland Operating	€12,000,000	1997	–	167 MWe	5–7	–

Since biomass has a high concentration of these elements and a substantial quantity of oxygen, it is believed to be even more reactive than coal [87] after its char is continually destroyed during gasification, leaving only trace residues. Consequently, co-gasification with other fuels is anticipated to gain from these characteristics. Some writers emphasize that alkali elements exacerbate buildup, particularly when chlorine is also present [73], jeopardizing the gasifier's operability, although countermeasures have previously been created [73].

Daniel Pio et al. [88] showed that the process is stable and that RDF and biomass work well together. This showed that co-gasification produces better gasification products than processes that only use biomass. During the experiments, no slag, clumping, or loss of fluidity was seen. RDF raised the concentrations of CH_4 and C_2H_4 significantly, while CO concentrations went down. This could be because of the thermal cracking of the plastic polymers in the RDF pellets and the catalytic effect of the ashes from the RDF pellets, which are high in alkali and alkali earth metal. No significant trends were seen in how the H_2 concentration changed with the RDF addition. This suggests that the ER and bed temperature may have a more significant effect on the production of this gaseous compound.

Still, because the CO concentration went down as the RDF weight percentage went up, higher H_2:CO molar ratios were found. In terms of efficiency parameters, the LHV of the producer gas went up as the RDF weight percentage increased. This was mainly because the concentrations of CH_4 and C_2H_4 increased. This happened more often in experiments where the ER was higher. Y_{gas} values also increased a little when the RDF weight percentage increased. But this effect could be hidden by changes in the ER that happen on their own and have a significant effect on the Y_{gas}.

Regarding cold gas efficiency (CGE) and carbon conversion efficiency (CCE), the RDF weight percentage increase seems to have a positive effect, but this is not clear because there are different effects. In terms of CGE, when the RDF weight percentage went up, the producer gas LHV and Y_{gas} went up, which led to a higher CGE. However, because the RDF has a higher LHV than the pine pellets and chips, an increase in the RDF weight percentage also means an increase in the energy content of the feedstock mixture, which means that the producer gas LHV and Y_{gas} have to go up to keep the process running efficiently. Regarding CCE, the higher weight percentage of RDF led to higher concentrations of CH_4 and C_2H_4, which led to higher CCE values. However, the higher weight percentage of RDF also led to lower CO concentrations, leading to lower CCE values. So, this information helps with the scaling up and commercialization of RDF gasification technologies and encourages more research on this topic. Due to the high availability and low cost of waste, adding RDF to the feedstock mix for gasification plants could significantly improve the economic viability and environmental benefits of future gasification plants. For example, this needs to be looked at correctly in future work using an integrated technoeconomic or life cycle analysis. Also, gaseous pollutants (like hydrochloric acid, sulfur oxides, and nitrogen species), particulate matter, the bottom bed, and volatile ashes must be characterized during RDF gasification and co-gasification with biomass to understand how this fuel affects gasification processes entirely. These things are essential for valorizing both municipal and biomass wastes, which could make waste management and energy supply more sustainable.

The use of RDF for WtE has advantages and disadvantages. Figure 4.4 provides a brief SWOT analysis of the RDF market, highlighting its strengths, weaknesses, opportunities, and threats. Consequently, modifying laws, rules, and objectives are the most viable options for mitigating climate change. Compared to RDF, the low price of oil and coal poses a significant challenge. This market is currently unregulated, but work is already underway to develop standards for its regulation, such as ISO/TC 300, which might stimulate this sector. Lastly, this research identifies emerging nations as prospective market-developing participants.

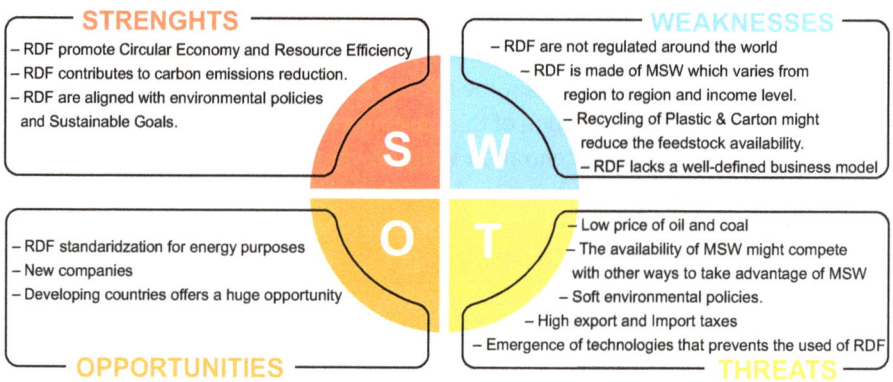

STRENGHTS
- RDF promote Circular Economy and Resource Efficiency
- RDF contributes to carbon emissions reduction.
- RDF are aligned with environmental policies and Sustainable Goals.

WEAKNESSES
- RDF are not regulated around the world
- RDF is made of MSW which varies from region to region and income level.
- Recycling of Plastic & Carton might reduce the feedstock availability.
- RDF lacks a well-defined business model

OPPORTUNITIES
- RDF standaridzation for energy purposes
- New companies
- Developing countries offers a huge opportunity

THREATS
- Low price of oil and coal
- The availability of MSW might compete with other ways to take advantage of MSW
- Soft environmental policies.
- High export and Import taxes
- Emergence of technologies that prevents the used of RDF

Figure 4.4: RDF SWOT analysis.

Large-scale urbanization has caused many problems, but those related to mobility, safety, health, well-being, sanitation, and the right way to handle MSW stand out. It is important to note that a waste energy recovery plant (WTE) is not a method of producing energy. Instead, it is a way to clean up the environment, and the energy it produces is a valuable by-product. This context is essential to show the authorities what WtE plants are and why they are important, especially in terms of cost and benefit compared to other ways to make electricity. Biomass and MSW could become crucial primary energy sources worldwide.

Still, a great deal needs to be accomplished. But the greatest urgency depends on real policy integration that makes it possible for all the different bioenergy actors to come together. So, promote the economic and environmental benefits that can come from pyrolysis and gasification of biomass.

References

[1]   J. A. M. Chavando, V. B. Silva, L. A. C. Tarelho, J. S. Cardoso, and D. Eusébio, "Snapshot review of refuse-derived fuels," *Utilities Policy*, vol. 74, p. 101316, Feb. 2022, doi: 10.1016/J.JUP.2021.101316.
[2]   World Energy Council, "Waste to Energy," 2013.
[3]   R. Gumisiriza, J. F. Hawumba, M. Okure, and O. Hensel, "Biomass waste-to-energy valorisation technologies: A review case for banana processing in Uganda," *Biotechnology for Biofuels*, 2017, vol. 10, no. 1, pp. 1–29, Jan. 2017, doi: 10.1186/S13068-016-0689-5.
[4]   A. Bosmans, I. Vanderreydt, D. Geysen, and L. Helsen, "The crucial role of Waste-to-Energy technologies in enhanced landfill mining: A technology review," *Journal of Cleaner Production*, vol. 55, pp. 10–23, Sep. 2013, doi: 10.1016/J.JCLEPRO.2012.05.032.
[5]   J. Alvarez, et al., "Hydrogen production from biomass and plastic mixtures by pyrolysis-gasification," *International Journal of Hydrogen Energy*, vol. 39, no. 21, pp. 10883–10891, Jul. 2014, doi: 10.1016/J.IJHYDENE.2014.04.189.
[6]   G. Ruoppolo, P. Ammendola, R. Chirone, and F. Miccio, "H2-rich syngas production by fluidized bed gasification of biomass and plastic fuel," *Waste Management*, vol. 32, no. 4, pp. 724–732, Apr. 2012, doi: 10.1016/J.WASMAN.2011.12.004.
[7]   F. Pinto, C. Franco, R. N. André, M. Miranda, I. Gulyurtlu, and I. Cabrita, "Co-gasification study of biomass mixed with plastic wastes," *Fuel*, vol. 81, no. 3, pp. 291–297, Feb. 2002, doi: 10.1016/S0016-2361(01)00164-8.
[8]   F. Pinto, et al., "Gasification improvement of a poor quality solid recovered fuel (SRF). Effect of using natural minerals and biomass wastes blends," *Fuel*, vol. 117, no. PARTB, pp. 1034–1044, Jan. 2014, doi: 10.1016/J.FUEL.2013.10.015.
[9]   M. L. Mastellone, L. Zaccariello, and U. Arena, "Co-gasification of coal, plastic waste and wood in a bubbling fluidized bed reactor," *Fuel*, vol. 89, no. 10, pp. 2991–3000, Oct. 2010, doi: 10.1016/J.FUEL.2010.05.019.
[10] I. I. Ahmed, N. Nipattummakul, and A. K. Gupta, "Characteristics of syngas from co-gasification of polyethylene and woodchips," *Applied Energy*, vol. 88, no. 1, pp. 165–174, Jan. 2011, doi: 10.1016/J.APENERGY.2010.07.007.
[11] M. Lapuerta, J. J. Hernández, A. Pazo, and J. López, "Gasification and co-gasification of biomass wastes: Effect of the biomass origin and the gasifier operating conditions," *Fuel Processing Technology*, vol. 89, no. 9, pp. 828–837, Sep. 2008, doi: 10.1016/J.FUPROC.2008.02.001.
[12] A. Ramos, E. Monteiro, V. Silva, and A. Rouboa, "Co-gasification and recent developments on waste-to-energy conversion: A review," *Renewable and Sustainable Energy Reviews*, vol. 81, Elsevier Ltd, pp. 380–398, 2018, doi: 10.1016/j.rser.2017.07.025.
[13] A. Thorenz, L. Wietschel, D. Stindt, and A. Tuma, "Assessment of agroforestry residue potentials for the bioeconomy in the European Union," *Journal of Cleaner Production*, vol. 176, pp. 348–359, Mar. 2018, doi: 10.1016/j.jclepro.2017.12.143.
[14] S. Panthapulakkal and M. Sain, The use of wheat straw fibres as reinforcements in composites. In *Biofiber Reinforcements in Composite Materials*, Elsevier Inc., 2015, pp. 423–453, doi: 10.1533/9781782421276.4.423.
[15] S. Arasaretnam and T. Kirudchayini, "Studies on synthesis, characterization of modified phenol formaldehyde resin and metal adsorption of modified resin derived from lignin biomass," *Emerging Science Journal*, vol. 3, no. 2, pp. 101–108, Apr. 2019, doi: 10.28991/ESJ-2019-01173.

[16] D. Kwabena Dadzie, A. K. Kaliluthin, and D. R. Kumar, "Civil engineering journal exploration of waste plastic bottles use in construction," vol. 6, no. 11, 2020, doi: 10.28991/cej-2020-03091616.

[17] U. Environmental Protection Agency, O. of Land, E. Management, and O. of Resource Conservation, "Advancing Sustainable Materials Management: 2017 Fact Sheet Assessing Trends in Material Generation, Recycling, Composting, Combustion with Energy Recovery and Landfilling in the United States," 2017.

[18] R. Avtar, S. Tripathi, A. K. Aggarwal, and P. Kumar, "Population-urbanization-energy nexus: A review," *Resources*, vol. 8, no. 3, pp. 1–21, 2019, doi: 10.3390/resources8030136.

[19] S. Kaza, L. Yao, P. Bhada-Tata, and F. van Woerden, *What A Waste 2.0: A Global Snapshot of Solid Waste Management to 2050*, The World Bank, 2018, doi: 10.1596/978-1-4648-1329-0.

[20] The World Bank, "World Bank Country and Lending Groups," 2020. https://datahelpdesk.worldbank.org/knowledgebase/articles/906519-world-bank-country-and-lending-groups (accessed May 05, 2021).

[21] S. Hemidat, M. Saidan, S. Al-Zu'bi, M. Irshidat, A. Nassour, and M. Nelles, "Potential utilization of RDF as an alternative fuel to be used in cement industry in Jordan," *Sustainability (Switzerland)*, vol. 11, no. 20, 2019, doi: 10.3390/su11205819.

[22] A. M. L. Násner, et al., "Refuse Derived Fuel (RDF) production and gasification in a pilot plant integrated with an Otto cycle ICE through Aspen plusTM modelling: Thermodynamic and economic viability," *Waste Management*, vol. 69, pp. 187–201, 2017, doi: 10.1016/j.wasman.2017.08.006.

[23] J. A. M. Chavando, V. Silva, D. R. D. S. Guerra, D. Eusébio, J. S. Cardoso, and L. A. C. Tarelho, "Review chapter: Waste to energy through pyrolysis and gasification in Brazil and Mexico," *Gasification [Working Title]*, Jun. 2021, doi: 10.5772/INTECHOPEN.98383.

[24] V. Singh Sikarwar, et al., "An overview of advances in biomass gasification," *Energy Environ. Sci*, vol. 9, p. 2939, 2016, doi: 10.1039/c6ee00935b.

[25] E. Slow, "RDF exports decline in 2018," Feb. 2019.

[26] FACT.MR, "Refuse-Derived Fuel Market Forecast, Trend Analysis & Competition Tracking – Global Market Insights 2019 – 2029," 2018. https://www.factmr.com/report/1461/refuse-derived-fuel-market (accessed Apr. 06, 2021).

[27] BusinessWire and Technavio, "Global Waste to Energy Market 2020-2024," 2020. https://www.businesswire.com/news/home/20200305005945/en/Global-Waste-to-Energy-Market-2020-2024-Evolving-Opportunities-with-Babcock-Wilcox-Enterprises-Inc.-and-China-Everbright-International-Ltd.-Technavio

[28] L. Makarichi, W. Jutidamrongphan, and K. Anan Techato, "The evolution of waste-to-energy incineration: A review," *Renewable and Sustainable Energy Reviews*, vol. 91, Elsevier Ltd, pp. 812–821, Aug. 2018, doi: 10.1016/j.rser.2018.04.088.

[29] M. J. Colinet, J. M. Cansino, J. M. González-Limón, and M. Ordóñez, "Toward a less natural gas dependent energy mix in Spain: Crowding-out effects of shifting to biomass power generation," *Utilities Policy*, vol. 31, pp. 29–35, Dec. 2014, doi: 10.1016/j.jup.2014.07.006.

[30] W. Y. Chen, T. Suzuki, and M. Lackner, *Handbook of Climate Change Mitigation and Adaptation*, second edition, vol. 1–4, Springer International Publishing, 2016, doi: 10.1007/978-3-319-14409-2.

[31] K. Isaac and S. O. Bada, "The co-combustion performance and reaction kinetics of refuse derived fuels with South African high ash coal," *Heliyon*, vol. 6, no. 1, p. e03309, Jan. 2020, doi: 10.1016/j.heliyon.2020.e03309.

[32] A. Szydełko, W. Ferens, and W. Rybak, "The effect of mineral additives on the process of chlorine bonding during combustion and co-combustion of solid recovered fuels," *Waste Management*, vol. 102, pp. 624–634, Feb. 2020, doi: 10.1016/j.wasman.2019.10.032.

[33] S. Zhang, X. Lin, Z. Chen, X. Li, X. Jiang, and J. Yan, "Influence on gaseous pollutants emissions and fly ash characteristics from co-combustion of municipal solid waste and coal by a drop tube furnace," *Waste Management*, vol. 81, pp. 33–40, Nov. 2018, doi: 10.1016/ j.wasman.2018.09.048.

[34] A. Sever Akdağ, A. Atimtay, and F. D. Sanin, "Comparison of fuel value and combustion characteristics of two different RDF samples," *Waste Management*, vol. 47, pp. 217–224, Jan. 2016, doi: 10.1016/j.wasman.2015.08.037.

[35] F. Duan, J. Liu, C. S. Chyang, C. H. Hu, and J. Tso, "Combustion behavior and pollutant emission characteristics of RDF (refuse derived fuel) and sawdust in a vortexing fluidized bed combustor," *Energy*, vol. 57, pp. 421–426, Aug. 2013, doi: 10.1016/j.energy.2013.04.070.

[36] H. Wu, P. Glarborg, F. J. Frandsen, K. Dam-Johansen, P. A. Jensen, and B. Sander, "Co-combustion of pulverized coal and solid recovered fuel in an entrained flow reactor – General combustion and ash behaviour," *Fuel*, vol. 90, no. 5, pp. 1980–1991, May. 2011, doi: 10.1016/ j.fuel.2011.01.037.

[37] K. Whiting, S. Wood, and M. Fanning, "Waste Technologies: Waste to Energy Facilities A Report for the Strategic Waste Infrastructure Planning (SWIP) Working Group, commissioned by the Government of Western," 2013.

[38] C. S. Psomopoulos, "Residue Derived Fuels as an Alternative Fuel for the Hellenic Power Generation Sector and their Potential for Emissions Reduction," Sep. 2014, doi: 10.3934/ energy.2014.3.321.

[39] F. Gao, *Pyrolysis of Waste Plastics into Fuels*. University of Canterbury, 2010.

[40] C. Z. Zaman, et al., Pyrolysis: A sustainable way to generate energy from waste. In *Pyrolysis*, InTech, 2017, doi: 10.5772/intechopen.69036.

[41] N. Grioui, K. Halouani, and F. A. Agblevor, "Assessment of upgrading ability and limitations of slow co-pyrolysis: Case of olive mill wastewater sludge/waste tires slow co-pyrolysis," *Waste Management*, vol. 92, pp. 75–88, Jun. 2019, doi: 10.1016/j.wasman.2019.05.016.

[42] S. Sfakiotakis and D. Vamvuka, "Study of co-pyrolysis of olive kernel with waste biomass using TGA/DTG/MS," *Thermochimica Acta*, vol. 670, pp. 44–54, Dec. 2018, doi: 10.1016/ j.tca.2018.10.006.

[43] R. Miandad, M. A. Barakat, A. S. Aburiazaiza, M. Rehan, I. M. I. Ismail, and A. S. Nizami, "Effect of plastic waste types on pyrolysis liquid oil," *International Biodeterioration & Biodegradation*, vol. 119, pp. 239–252, Apr. 2017, doi: 10.1016/j.ibiod.2016.09.017.

[44] H. E. Whyte, K. Loubar, S. Awad, and M. Tazerout, "Pyrolytic oil production by catalytic pyrolysis of refuse-derived fuels: Investigation of low cost catalysts," *Fuel Processing Technology*, vol. 140, pp. 32–38, Dec. 2015, doi: 10.1016/j.fuproc.2015.08.022.

[45] J. D. Martínez, et al., "Co-pyrolysis of biomass with waste tyres: Upgrading of liquid bio-fuel," *Fuel Processing Technology*, vol. 119, pp. 263–271, Mar. 2014, doi: 10.1016/ j.fuproc.2013.11.015.

[46] I. H. Hwang, J. Kobayashi, and K. Kawamoto, "Characterization of products obtained from pyrolysis and steam gasification of wood waste, RDF, and RPF," *Waste Management*, vol. 34, no. 2, pp. 402–410, Feb. 2014, doi: 10.1016/j.wasman.2013.10.009.

[47] J. Eke, J. A. Onwudili, and A. V. Bridgwater, "Influence of moisture contents on the fast pyrolysis of trommel fines in a bubbling fluidized bed reactor," *Waste and Biomass Valorization*, vol. 11, pp. 3711–3722, 2020, doi: 10.1007/s12649-018-00560-2.

[48] F. Sembiring, C. W. Purnomo, and S. Purwono, "Catalytic pyrolysis of waste plastic mixture," *IOP Conference Series: Materials Science and Engineering*, vol. 316, no. 1, p. 012020, Mar, 2018, doi: 10.1088/1757-899X/316/1/012020.

[49] J. A. M. Chavando, E. C. J. de Matos, V. B. Silva, L. A. C. Tarelho, and J. S. Cardoso, "Pyrolysis characteristics of RDF and HPDE blends with biomass," *International Journal of Hydrogen Energy*, Dec. 2021, doi: 10.1016/J.IJHYDENE.2021.11.062.

[50] M. Åhman and L. J. Nilsson, "Path dependency and the future of advanced vehicles and biofuels," *Utilities Policy*, vol. 16, no. 2, pp. 80–89, Jun. 2008, doi: 10.1016/j.jup.2007.11.003.

[51] D. T. Pio, L. A. C. Tarelho, A. M. A. Tavares, M. A. A. Matos, and V. Silva, "Co-gasification of refused derived fuel and biomass in a pilot-scale bubbling fluidized bed reactor," vol. 206, October 2019, 2020, doi: 10.1016/j.enconman.2020.112476.

[52] C. Nobre, A. Longo, C. Vilarinho, and M. Gonçalves, "Gasification of pellets produced from blends of biomass wastes and refuse derived fuel chars," *Renewable Energy*, Mar. 2020, doi: 10.1016/j.renene.2020.03.077.

[53] P. Šuhaj, J. Haydary, J. Husár, P. Steltenpohl, and I. Šupa, "Catalytic gasification of refuse-derived fuel in a two-stage laboratory scale pyrolysis/gasification unit with catalyst based on clay minerals," *Waste Management*, vol. 85, pp. 1–10, Feb. 2019, doi: 10.1016/j.wasman.2018.11.047.

[54] N. Indrawan, S. Thapa, P. R. Bhoi, R. L. Huhnke, and A. Kumar, "Electricity power generation from co-gasification of municipal solid wastes and biomass: Generation and emission performance," *Energy*, vol. 162, pp. 764–775, Nov. 2018, doi: 10.1016/j.energy.2018.07.169.

[55] T. Khosasaeng and R. Suntivarakorn, Effect of equivalence ratio on an efficiency of single throat downdraft gasifier using RDF from municipal solid waste. In *Energy Procedia*, vol. 138, pp. 784–788, Oct. 2017, doi: 10.1016/j.egypro.2017.10.066.

[56] T. Robinson, B. Bronson, P. Gogolek, and P. Mehrani, "Comparison of the air-blown bubbling fluidized bed gasification of wood and wood-PET pellets," *Fuel*, vol. 178, pp. 263–271, Aug. 2016, doi: 10.1016/j.fuel.2016.03.038.

[57] Z. Ong, et al., "Co-gasification of woody biomass and sewage sludge in a fixed-bed downdraft gasifier," *AIChE Journal*, vol. 61, no. 8, pp. 2508–2521, Aug. 2015, doi: 10.1002/aic.14836.

[58] R. A. Moghadam, S. Yusup, Y. Uemura, B. L. F. Chin, H. L. Lam, and A. Al Shoaibi, "Syngas production from palm kernel shell and polyethylene waste blend in fluidized bed catalytic steam co-gasification process," *Energy*, vol. 75, pp. 40–44, Oct. 2014, doi: 10.1016/j.energy.2014.04.062.

[59] P. Straka and O. Bičáková, "Hydrogen-rich gas as a product of two-stage co-gasification of lignite/waste plastics mixtures," *International Journal of Hydrogen Energy*, vol. 39, no. 21, pp. 10987–10995, Jul. 2014, doi: 10.1016/j.ijhydene.2014.05.054.

[60] P. Lahijani, Z. A. Zainal, A. R. Mohamed, and M. Mohammadi, "Co-gasification of tire and biomass for enhancement of tire-char reactivity in CO_2 gasification process," *Bioresource Technology*, vol. 138, pp. 124–130, Jun. 2013, doi: 10.1016/j.biortech.2013.03.179.

[61] L. Peng, Y. Wang, Z. Lei, and G. Cheng, "Co-gasification of wet sewage sludge and forestry waste in situ steam agent," *Bioresource Technology*, vol. 114, pp. 698–702, Jun. 2012, doi: 10.1016/j.biortech.2012.03.079.

[62] M. Seggiani, M. Puccini, G. Raggio, and S. Vitolo, "Effect of sewage sludge content on gas quality and solid residues produced by cogasification in an updraft gasifier," *Waste Management*, vol. 32, no. 10, pp. 1826–1834, Oct. 2012, doi: 10.1016/j.wasman.2012.04.018.

[63] G. Ruoppolo, P. Ammendola, R. Chirone, and F. Miccio, "H 2-rich syngas production by fluidized bed gasification of biomass and plastic fuel," *Waste Management*, vol. 32, no. 4, pp. 724–732, Apr. 2012, doi: 10.1016/j.wasman.2011.12.004.

[64] I. I. Ahmed, N. Nipattummakul, and A. K. Gupta, "Characteristics of syngas from co-gasification of polyethylene and woodchips," *Applied Energy*, vol. 88, no. 1, pp. 165–174, Jan. 2011, doi: 10.1016/j.apenergy.2010.07.007.

[65] D. D. Wall, "Technical and Economical Evaluation of selected Processes for Chemical Syngas Generation," Montanuniversität Leoben, 2012.

[66] V. Litvinenko and B. Meyer, "Syngas utilization technologies," *Syngas Production: Status and Potential for Implementation in Russian Industry*, pp. 23–46, 2018, doi: 10.1007/978-3-319-70963-5_5.

[67] H. Li, D. Mehmood, E. Thorin, and Z. Yu, "ScienceDirect biomethane production via anaerobic digestion and biomass gasification selection and/or peer-review under responsibility of ICAE," *Energy Procedia*, vol. 105, pp. 1172–1177, 2017, doi: 10.1016/j.egypro.2017.03.490.

[68] G. Evans and C. Smith, "Biomass to liquids technology," *Comprehensive Renewable Energy*, vol. 5, pp. 155–204, Jan. 2012, doi: 10.1016/B978-0-08-087872-0.00515-1.

[69] European Commission and Cordis, "Combustion of Rdf Produced from Selected Industrial Waste, for the Production of Heat and Electricity," 2020. https://cordis.europa.eu/project/id/BM.-00295-86

[70] CORDIS and EC, "Polygeneration through gasification utilising secondary fuels derived from MSW," 2015. https://cordis.europa.eu/project/id/219062/fr (accessed May 05, 2021).

[71] RICARDO-AEA, "Case Study 1. Lahti Gasification Facility, Finland," Jul. 2013.

[72] M. Siedlecki, W. de Jong, and A. H. M. Verkooijen, "Fluidized bed gasification as a mature and reliable technology for the production of bio-syngas and applied in the production of liquid transportation fuels – a review," *Energies*, 2011, vol. 4, no. 3, pp. 389–434, doi: 10.3390/EN4030389.

[73] A. A. Khan, W. de Jong, P. J. Jansens, and H. Spliethoff, "Biomass combustion in fluidized bed boilers: Potential problems and remedies," *Fuel Processing Technology*, vol. 90, no. 1, pp. 21–50, Jan. 2009, doi: 10.1016/J.FUPROC.2008.07.012.

[74] L. Devi, K. J. Ptasinski, and F. J. J. G. Janssen, "A review of the primary measures for tar elimination in biomass gasification processes," *Biomass and Bioenergy*, vol. 24, no. 2, pp. 125–140, Feb. 2003, doi: 10.1016/S0961-9534(02)00102-2.

[75] S. Rapagnà, N. Jand, A. Kiennemann, and P. U. Foscolo, "Steam-gasification of biomass in a fluidised-bed of olivine particles," *Biomass and Bioenergy*, vol. 19, no. 3, pp. 187–197, Sep. 2000, doi: 10.1016/S0961-9534(00)00031-3.

[76] M. Siedlecki and W. de Jong, "Biomass gasification as the first hot step in clean syngas production process – Gas quality optimization and primary tar reduction measures in a 100 kW thermal input steam–oxygen blown CFB gasifier," *Biomass and Bioenergy*, vol. 35, no. SUPPL. 1, pp. S40–S62, Oct. 2011, doi: 10.1016/J.BIOMBIOE.2011.05.033.

[77] M. A. Hazrat, M. G. Rasul, and M. M. K. Khan, "A study on thermo-catalytic degradation for production of clean transport fuel and reducing plastic wastes," *Procedia Engineering*, vol. 105, pp. 865–876, Jan. 2015, doi: 10.1016/J.PROENG.2015.05.108.

[78] J. M. de Andrés, A. Narros, and M. E. Rodríguez, "Behaviour of dolomite, olivine and alumina as primary catalysts in air–steam gasification of sewage sludge," *Fuel*, vol. 90, no. 2, pp. 521–527, Feb. 2011, doi: 10.1016/J.FUEL.2010.09.043.

[79] T. Wigmans, J. C. Göebel, and J. A. Moulijn, "The influence of pretreatment conditions on the activity and stability of sodium and potassium catalysts in carbon-steam reactions," *Carbon N Y*, vol. 21, no. 3, pp. 295–301, Jan. 1983, doi: 10.1016/0008-6223(83)90094-5.

[80] M. J. Veraa and A. T. Bell, "Effect of alkali metal catalysts on gasification of coal char," *Fuel*, vol. 57, no. 4, pp. 194–200, Apr. 1978, doi: 10.1016/0016-2361(78)90116-3.

[81] S. V. Vassilev, D. Baxter, L. K. Andersen, and C. G. Vassileva, "An overview of the chemical composition of biomass," *Fuel*, vol. 89, no. 5, pp. 913–933, May. 2010, doi: 10.1016/ J.FUEL.2009.10.022.

[82] S. Krerkkaiwan, C. Fushimi, A. Tsutsumi, and P. Kuchonthara, "Synergetic effect during co-pyrolysis/gasification of biomass and sub-bituminous coal," *Fuel Processing Technology*, vol. 115, pp. 11–18, Nov. 2013, doi: 10.1016/J.FUPROC.2013.03.044.

[83] P. Lahijani, Z. A. Zainal, A. R. Mohamed, and M. Mohammadi, "CO2 gasification reactivity of biomass char: Catalytic influence of alkali, alkaline earth and transition metal salts," *Bioresource Technology*, vol. 144, pp. 288–295, Sep. 2013, doi: 10.1016/ J.BIORTECH.2013.06.059.

[84] P. McKendry, "Energy production from biomass (part 1): Overview of biomass," *Bioresource Technology*, vol. 83, no. 1, pp. 37–46, May. 2002, doi: 10.1016/S0960-8524(01)00118-3.

[85] F. A. López, T. A. Centeno, F. J. Alguacil, B. Lobato, A. López-Delgado, and J. Fermoso, "Gasification of the char derived from distillation of granulated scrap tyres," *Waste Management*, vol. 32, no. 4, pp. 743–752, Apr. 2012, doi: 10.1016/J.WASMAN.2011.08.006.

[86] Y. Lin, X. Ma, Z. Yu, and Y. Cao, "Investigation on thermochemical behavior of co-pyrolysis between oil-palm solid wastes and paper sludge," *Bioresource Technology*, vol. 166, pp. 444–450, Aug. 2014, doi: 10.1016/J.BIORTECH.2014.05.101.

[87] C. Brage, Q. Yu, G. Chen, and K. Sjöström, "Tar evolution profiles obtained from gasification of biomass and coal," *Biomass and Bioenergy*, vol. 18, no. 1, pp. 87–91, Jan. 2000, doi: 10.1016/S0961-9534(99)00069-0.

[88] D. T. Pio, L. A. C. Tarelho, A. M. A. Tavares, M. A. A. Matos, and V. Silva, "Co-gasification of refused derived fuel and biomass in a pilot-scale bubbling fluidized bed reactor," *Energy Conversion and Management*, vol. 206, p. 112476, Feb. 2020, doi: 10.1016/ j.enconman.2020.112476.

5 Technoeconomic analysis of a biomass gasification power plant

The electricity production from biomass sources is expanding worldwide, accounting for 3600 biomass power plants with a total capacity of over 51 GW [1].

Europe takes the lead in this segment, accounting for more than 1000 active biomass power plants alone, due in part to the dense woodlands of Scandinavian countries and the continuing and comprehensive subsidization granted for power plants in countries such as Germany [2].

The economic and energetic performance of a power plant depends on many variables, namely, woody biomass availability, location, cost of operation and maintenance, plant capacity, logistics, environmental benefits, and financial incentives, to mention a few. In this context, to successfully assess the viability of a biomass energy system, a strategic analysis, such as a technoeconomic and environmental study, must be conducted preceding deployment to evaluate the feasibility and sustainability of the projects, as also to identify possible variables hampering its success.

The fuel input is also a subject of major concern when performing gasification processes for electricity production as the biomass fuel demands required to keep the system fully functional without any electricity production disruption is considerable [3]. Thus, performing gasification with forestry biomass blends instead of a single biomass gasification method is advantageous once it enhances feedstock availability by considering alternative fuel sources for the power plant. Such practice allows to maintain a stable biomass supply, compensate for eventual seasonal influences in the feedstock availability, avert storage issues and high transportation costs, and avoid disruption by providing supplementary resource options [4].

In this chapter, the authors deliver a forest biomass blends gasification power plant technoeconomic analysis in a forest located in mainland Portugal. First, an investment analysis was performed by calculating the net present value (NPV), internal rate of return (IRR), and payback period (PBP). Then, performance analysis and uncertainties of the applied economic model calculations were determined by coupling a Monte Carlo sensitivity analysis.

5.1 Methodology

Following an open public mindset, this chapter considers the construction and operation of a biomass gasification power plant dealing essentially with forest residue blends generated in the central region of Portugal (Figure 5.1a). In an attempt to deliver an approach as close as possible to a real scenario, this analysis was built

https://doi.org/10.1515/9783110758214-005

based on the existing literature review and evaluation reports (both national and international) concerning investment projects in forest biomass power plants [5–11]. Table 5.1 comprises the main technical and operational characteristics of the designed unit.

Table 5.1: Summary of the gasification power plant's main characteristics.

Technical and operational baselines of the gasification power plant investment project		References
Location	Coimbra District, Portugal	–
Installed power (MW)	11	[6, 12]
Project lifetime (years)	25	[6]
Plant availability (days) (including 10 scheduled shutdown days)	365	[6, 9]
Total area of land occupied by the unit (ha) (includes buildings and biomass storage treatment park)	6	[6]
Biomass average consumption (tons/year) (calculated assuming a standard humidity degree of 40%)	115,500	[6, 12]
Biomass cost (€/ton) (in 2018, it includes transportation expenses)	35	[6]
Area of influence (radius in km) (collection radius of forest biomass)	25	[6]
Electricity production (MWh/year) (at cruising speed, accounting for powerline losses)	78,436.72	[6]
Sales of electricity (€/MWh) (tariff charged in 2018)	121.34	[13]
System's overall efficiency (%)	30	[14]
Powerline losses (%)	1.8	[6]
Number of workers	16	[6]
Gasification power plant's main components	Biomass storage and treatment park Biomass bubbling fluidized bed gasifier Gas cooling and cleaning system Gas turbine and generator Intern heat recovery system Electrical substation Control room	[6, 8, 9, 14]

The geographical forest biomass availability and its supply area are depicted in Figure 5.1b. The preferred location to lodge the project in view was chosen given the higher biomass availability in the central Portuguese region, carrying a maximum estimated biomass availability of 122,282 tons/year in the densest forest regions, better ensuring the biomass supply by already predicting potential future feedstock competition with other forestry industry sectors.

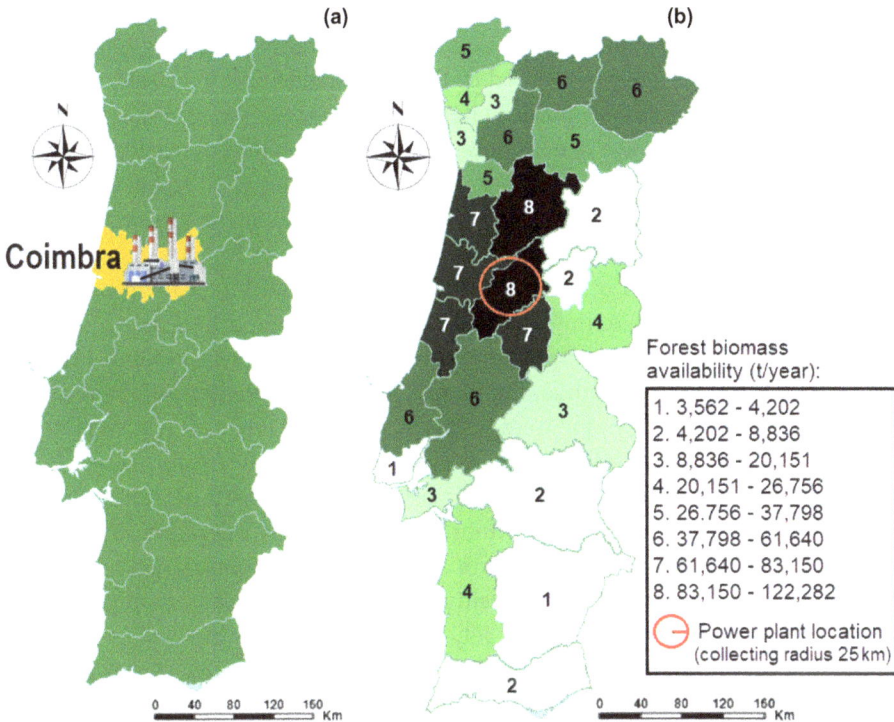

Figure 5.1: (a) Map of Portugal describing the location of the projected biomass gasification power plant; (b) forest biomass availability at regional scale and biomass supply area within the gasification power plant's 25 km collecting radius [15].

The biomass power plant project here considered extends to a total period of 27 years, from 2018 to 2045. This timeline comprises two major stages of development: the design and construction phase (2 years) and the exploration phase (25 years). The exploration phase is assumed to start in 2020. Plant availability of 365 days per year is assumed, with 10 scheduled shutdown days for maintenance operations.

Forest biomass harvesting and collection is assumed to occur mainly by cut-and-chip process on-site, followed by direct transportation to the power station for storage and processing. At the power plant, biomass processing integrates multiple sieves to ensure that the fuel meets the granulometry and uniformity required to feed into the gasifier. The electricity is produced by burning the biomass blends in a bubbling fluidized bed gasifier to generate gas. The produced gas delivers a combustible mixture composed of carbon monoxide (14.0–21.4%), hydrogen (2.0–12.7%), carbon dioxide (14.2–17.5%), methane (3.6–5.8%) and nitrogen (48.9–61.1%) [16]. The gas is then carried through a gas cooling and cleaning system before being channeled to a gas turbine connected to a generator which produces electricity. Further technical and process details regarding the biomass gasification process developed on an identical system can be found in [17]. Finally, a medium voltage connection line (60 kV) delivers the electricity to the national electrical grid up to 11 MW, according to the maximum installed power capacity licensed by the public tender. All the electrical energy exported to the substation, acknowledging powerline losses, is assumed to be sold to the national grid. The power line distance from the plant to the connection point will be at most 15 km. In this study, no heat sales were considered. All residual heat produced during the process is assumed to be recovered by an intern heat recovery system for cooling or heating purposes.

The power plant will operate by consuming residual forest biomass blends up to a minimum of 90% of the total fuel burned. However, other timber industry wastes or agriculture wastes may also complement the forestry blend primary fuel. Within its collection radius, the plant's provisioning policy will be directed preferentially to blends of pine trees (*Pinus pinaster* and *Pinus pinea*), eucalyptus (*Eucalyptus globulus*), and some cork oak (*Quercus rotundifolia*); areas with higher wildfire hazard; and by employing a center-periphery perspective, with impact on transport costs and profitability of the supply operation. The placement of the power plant project within a highly dense biomass region such as the central Portuguese region will positively impact the biomass supply chain as the transportation costs will highly rely on the availability of suitable biomass within the geographical area to be sold to the processing biomass power plant [18]. Pine trees and eucalyptus will compose most of the mixture due to their considerable prevalence in the plant's area of influence compared to cork oak and other forestry and agriculture residues [15]. The employed forestry biomass blends are within the acceptable performance range and do not hamper the gasifying unit's overall efficiency (around 30%, according to the manufacturer) or the obtained gas quality [14, 19–21]. Moreover, the research group has already studied the system performance behavior in dealing with forest biomass residues in an identical bubbling fluidized bed gasifier system at a pilot scale developed by the same manufacturer [17, 22].

According to the World Bank Group guidelines [11], to assure the supply of forest biomass to the plant during the start-up phase, a previously contracted supply plan scenario established with local forest producers and associations is assumed. Attending to the installed power and the predicted operative needs of the power plant, an estimated total of 40,646 ha of forest area, ensuring the collecting and selling of a predicted amount of 40,646 tons/year, is postulated. However, this precontracted biomass amount makes less than 50% of the requirements to operate the power plant at cruising speed, assumed to be around 115,500 tons/year. Thus, it is primordial to correctly estimate the future biomass needs of the plant to avoid disruption. Therefore, the average consumption of 10.5 kilotons/MW of installed capacity is employed. Biomass requirement calculations and cost per ton consider a base humidity of 40% as standard. This foreseeable biomass consumption is made after the completion of the projected investments. A strategic partnership with forest producers and associations gathers great importance given their direct influence on the development of biomass and forest management, ensuring the future supply of the plant and promoting increased availability of forest biomass in the region. Given these assumptions, it may be reasonable to ponder in the near future, in addition to the gasification of forestry residue blends, the co-gasification of forest biomass blends with municipal solid waste (or other sorts of solid residues) as a clever strategy to reduce exploration costs, increase plant's production efficiency, and avoid biomass exploration excess and consequent disequilibrium of ecosystems.

The personnel structure required to operate the power plant is assumed to be composed of 16 full-time employees, divided into 5 pay grades, 2 for management, 2 for specialized workers, 6 for rank and file, 2 for administrative, and 4 for surveillance [6].

In the economic analysis, costs related to each project stage are estimated and applied. Table 5.2 presents the input financial data and details the annually considered cost factors practiced to model the plant's technoeconomic feasibility. The cash flow cost and revenue analysis incurred in:

i.   an initial investment period related to the design and construction phase with 30% equity capital and 70% borrowed capital;
ii. amortizations of the debt contracted with the acquisition of the project;
iii. investments in fixed assets and working capital, which will be deducted to amortizations so to recover the investments;
iv. financial income from capital investments;
v.   costs related to operation and maintenance (O&M), employees, and structure; and
vi. revenues from selling electricity to the grid.

Cash flows for costs and revenue calculations along the project lifetime are applied to a spreadsheet-based economic model developed to calculate the NPV, IRR, and PBP. Details on the formulation of these methods are provided in the following section. Unlike the initial investment period, which only extends throughout the design and construction phase, from 2018 to 2019, all the remaining cash flows are extended for 25 years. Initially, all costs and revenues are updated to the year they are related to. Then, the total annual cash flow is calculated by summing up all the abovementioned costs and revenues for each year. Next, annual revenues are calculated by multiplying the annual electricity production of the plant, accounting for powerline losses, by the applicable selling price of electricity in the same year. Next, the annual cash flow is determined by balancing the total annual costs (outflows) and revenues

Table 5.2: Initial input financial data and detailed considered cost factors for the gasification power plant project.

Input financial parameters	Value	Remarks	References
Discount rate (WACC) (%)	8.18	After-tax	[6]
Inflation rate (%)	1.6	The rate applied in 2020	[23]
Equity capital (30%) (M€)	11.190	Values applicable during the investment period. Includes costs of bid submission, company formation expenses, due diligence, credit opening, land acquisition, plant and electric power line construction, equipment with a life cycle equal to the plant, and initial fixed assets investments	[6]
Borrowed capital (70%) (M€)	26.110		[6]
Amortizations (M€)	3.437	Starting value in 2020. Includes the regular debt payment throughout the project's lifetime	[6]
Fixed assets investments (M€)	0.305	Starting value in 2021 (deducted to amortizations)	[6]
Working capital investments (M€)	2.599	Starting value in 2020 (deducted to amortizations)	[6]
Financial income (M€)	0.009	Income during the investment period	[6]
O&M cost (M€)	5.149	Starting value in 2020. Includes consumption cost with biomass supply and transportation, electricity, piped water, ash transport and disposal, and maintenance of equipment and buildings	[6]
Employees' cost (M€)	0.454	Starting value in 2020. Includes wages, benefits, insurance expenses, and employers' social security charges according to the Portuguese remuneration table	[6]
Structure cost (M€)	0.307	Starting value in 2020. Includes administrative costs, consulting, marketing and communication, cleanings, training programs, travel and stays, service cars, and technical assistance	[6]
Total annual cost (M€)	6.748	In 2020	–
Output financial parameters			
Annual revenue (M€/year) (electricity selling)	10.07	In 2022 (cruising speed achieved)	–

(inflows). From this, a discounted cash flow analysis is developed by dividing the annual cash flow in each year by one plus the calculated discount rate after tax elevated to the relating year. This discount rate corresponded to a weighted average cost of all capital invested in the project, equity, and borrowed, known as the weighted average cost of capital (WACC). The WACC calculation method is described hereinafter. Finally, a cumulative NPV gives out the present worth of negative and positive investment cash flow.

For simplification purposes, equipment replacement investments necessary to restore installed capacity and maintain the plant's operating conditions are included within O&M costs cash flow. Also, it is assumed that all expenses occur at the end of the year to which they are related. All the analysis is carried out at current prices, revenues, and value-added tax rates. The applied inflation rates for 2019 and 2020 are based on the forecasts of the Bank of Portugal [23]. From 2021 onward, the inflation rate is considered the average for the last 10 years. Applied tariffs and price evolution throughout the project's lifetime are updated according to the consumer price index from the National Statistical Institute of Portugal [24]. Amortization rates are in accordance with the straight-line method. Tax on profit calculation accounts for tax incentives to the interiority, according to the Portuguese Law no. 171/99 *(Diário da República, 1999)*, given the beneficiary region in which the designed power plant is located. The interest rates considered result from the quotations provided by Portuguese banking for this type of project [6].

5.2 Economic model development

To develop the economy, three common approaches are combined to evaluate the biomass gasification power plant project viability over a lifetime period of 25 years: NPV, IRR, and PBP. Each of these methods carries its strengths and limitations. For instance, NPV allows keeping a good track of the cash flows over a given period. However, it is sensitive to discount rates, as a small increase or decrease in the discount rate will strongly affect the NPV final output. On the other hand, despite providing a more straightforward approach, IRR evaluates every investment by delivering merely one discount rate. As for PBP, it presents an easily perceptible and straightforward analysis when calculating the speed of return, yet, it considers no inflation, financing, or risks associated with the investment. In short, it goes without question that each method has its intended purpose. Thus, this analysis aims to deliver a strengthened economic evaluation toward an improved investment decision by combining these three methods.

5.2.1 Net present value

The NPV is an indicator used to evaluate the profitability of investment projects by summing all inflows and outflows of cash over the project lifetime. A positive NPV indicates that the project earnings surpass the expected costs, meaning that the project is profitable, while a negative NPV will event in a net loss. To assess the value of the project, financed by a combination of debt and equity, the discount rate applied to the cash flows in NPV calculation corresponds to an overall cost of capital from a weighted average cost of all capital sources invested in the project. The NPV considers this discount rate (calculated as WACC) over the project lifetime, giving the annual cash flows in present values. Hence, the NPV can be equated as follows [11]:

$$NPV(i,N) = \sum_{t=0}^{N} \frac{C_t}{(1+i)^t} \tag{5.1}$$

where i is the financial discount rate (WACC), C_t is the annual cash flow in the tth year, and N is the total number of years. Time period $t = 0$ relates to the investment during the design and construction phase.

The WACC is calculated based on the cost of equity and the cost of debt (borrowed capital), which allows determining the discount rate applicable to cash flows to calculate the NPV. The WACC is given by [6]

$$WACC = \left(\frac{E}{E+D}\right)K_e + \left(\frac{D}{E+D}\right)K_d(1-T) \tag{5.2}$$

where E is the market value of the equity, D is the market value of the debt, K_e is the cost of equity, K_d is the cost of debt, and T is the marginal tax rate.

5.2.2 Internal rate of return

The IRR stands as an indicator used to measure the profitability of an investment project. The higher the IRR, the greater the project's profitability will be. In addition, the IRR is the discount rate that makes the NPV of all cash flows equal to zero, determining the minimum rate of return to make the project viable. Therefore, the project is feasible if the IRR is higher than the discount rate (WACC). The IRR is calculated by [25]

$$NPV(IRR,N) = \sum_{t=0}^{N} \frac{C_t}{(1+IRR)^t} = 0 \tag{5.3}$$

5.2.3 Payback period

The PBP is the time required to reclaim the initial capital investments. The shorter the PBP, the stronger the project's financial feasibility. In this analysis, PBP is calculated by finding the year in which the cumulative NPV cash flow becomes positive and is formulated as follows [11]:

$$PBP = A + \frac{B}{C} \tag{5.4}$$

where A is the last year with a negative cumulative NPV, B is the absolute value of cumulative NPV at the end of year A, and C is the total annual cash flow during the year after A.

5.3 Economic model results and discussion

The economic analysis gives out the financial feasibility from an investor prospect of the project along its lifetime. Figure 5.2 shows the economic model results for the NPV, IRR, and PBP calculations. The NPV profile calculations are presented at various discount rates, from 0% to 24%, determining whether the project is accepted or rejected. The acceptance criterion regarding this approach is if the NPV is higher than zero, the project is accepted and rejected if otherwise. A higher discount rate will increase the NPV calculation denomination (eq. (5.1)), resulting in a lower present value.

In contrast, a lower discount rate will cause NPV to increase. So, the NPV profile behavior is inversely proportional to the discount rate. This NPV hasty variation underlines its sensitivity to the discount rate. After-tax, the power plant project NPV at the applied discount rate of 8.18% is €2.367M. The IRR is the return rate offered by the project and is given by the moment at which the NPV equals zero. For this project, the calculated IRR rate is 8.66%. The PBP is the year in which the cumulative cash flow turns positive. Thus, 23.1 years is the calculated amount of time needed to restore the initial capital investments, meeting the 22-year period for amortization expenses coverage delivered by the project cash flow results.

Table 5.3 provides the financial indicators retrieved from the literature for an 11 MW combustion power plant project operating in similar conditions in mainland Portugal to place these results in comparison with the competing biomass combustion technology. Comparatively, the combustion system provides enhanced economic feasibility with an increased NPV of €4.878M, an IRR of 9.95%, and a PBP of 21 years. These numbers come with no surprise once despite the broadly known technical and environmental advantages of gasification compared to combustion systems. Nevertheless, gasification technology still faces several disadvantages, mostly related to higher capital costs. Therefore, research developments in gasification are important as they may improve the long-term outlook and potential market share for this technology [26].

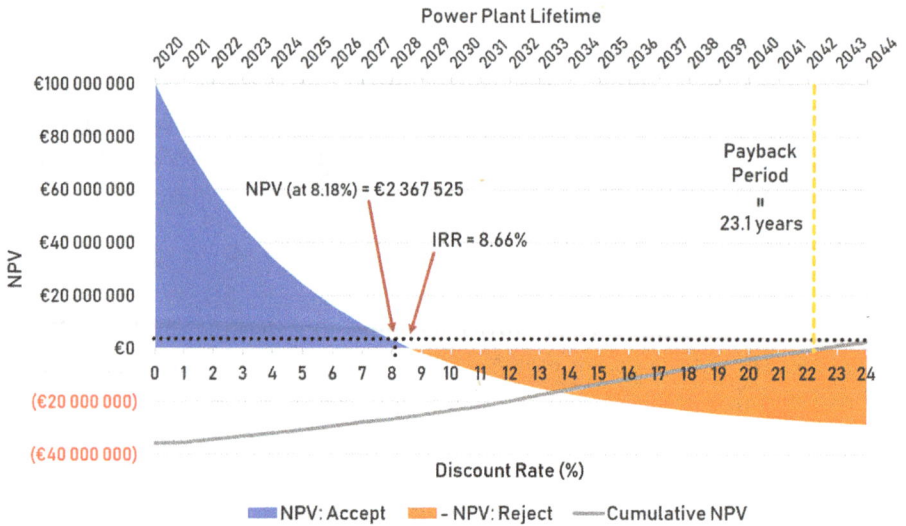

Figure 5.2: Financial indicators overview throughout the gasification power plant lifetime: NPV at a discount rate of 8.18% and IRR of the project. NPV profile variation (accept/reject) as a function of the discount rate and cumulative NPV throughout the power plant lifetime and its PBP.

Table 5.3: Financial indicators for an 11 MW biomass combustion technology power plant operating in similar conditions [6].

	Financial indicators		
	NPV	IRR	PBP
11 MW biomass combustion technology power plant	€4.787M	9.95%	21 years

Overall, a positive NPV, an IRR greater than the discount rate, and a PBP inferior to the power plant lifetime turn this project into an economically feasible investment from an economic point of view. However, despite its apparent feasibility, it may not be that economically attractive to most investors. Taking, for instance, a Portuguese example that occurred in 2006, in which tender releases from 15 biomass combustion power plants only 2 came to completion [3]. General failure was mainly attributed to poor location, high feedstock costs, supply chain, logistics, feedstock availability issues, the bureaucracy of tender procedures, and lack of bank financing support, which held back private investors. Factors that strongly affect the feasibility and effectiveness of the project influence decision-making. According to the World Bank Group, typical benchmarks for key financial parameters in biomass projects point out that the NPV ought to be a positive value, IRR must be greater than 10%, and PBP inferior to 10 years [11]. Obviously, these generalized criteria will differ by country risk and project-specific conditions. Yet, at first sight, these assumptions may compromise the economic feasibility of this project, discouraging investors less willing to risk. From this analysis, one can immediately assess the PBP and its relatively tight timeframe with the project lifespan, showing that it would require circa 92% of the project's lifetime to recoup the investment costs. Therefore, the longer the PBP, the less desirable the investment will be. This reinforces the importance of performing the economic analysis by comprising the three approaches and avoiding poor investment decisions. On that account, one must consider several risks associated when performing an investment analysis in forest biomass projects. Therefore, a sensitivity analysis is to be performed to determine which variables are critical to the project's success.

5.4 Sensitivity analysis

5.4.1 Monte Carlo simulation

A sensitivity analysis is carried out to assess the most critical variables considered for the project's performance to measure the risks associated with the project. The viability of financial indicators is determined by measuring their elasticity and testing the project performance reaction to stressful scenarios, simultaneously imposing either a favorable or unfavorable evolution of several variables.

Some input variables comprise a higher range of uncertainty than others. The following are considered to be the most significant economic and performance variables affecting the viability of the project:

i.   The initial investment, given the initial large amounts of capital outflow from which the project needs to recover.
ii. The discount rate, once it influences the present value of future costs and benefits.
iii. Besides its importance for the expected energy revenue, electricity sales price holds a long-term uncertainty being subdued by political decisions and market fluctuations.
iv. Electricity production, as revenues rely upon a stable annual output.
v.   Biomass cost, considering it is the main supply for the entire production.

A risk model based on Monte Carlo simulation is implemented within the economic model spreadsheet for the sensitivity analysis. Monte Carlo allows performing the risk analysis of the project by simulating a range of possible outcomes for a number of scenarios, assessing decision-making over uncertainty. The five input variables' sensitivity bounds are defined as unfavorable/favorable by varying the baseline value up to a ±10% range, testing the price evolution scenario of the investment [6]. These inputs' uncertainties will translate over the financial indicators (NPV, IRR, and PBP) selected as the corresponding outputs in the risk model. The Monte Carlo simulation is conducted for a total of 10,000 iterations. The input variables are sampled for each iteration based on the defined sensitivity bounds range. All other variables within the economic model are maintained unchanged during this analysis. A triangular distribution is considered for each variable, requiring the input of a lower limit (unfavorable value), a most likely value (baseline value), and an upper limit (favorable value). This sort of distribution is often applied to risk analysis due to its relative mathematical simplicity while generating enough random samples to identify the most sensitive parameters [27]. Additional details concerning the Monte Carlo analysis can be found elsewhere [3, 28].

Figure 5.3 depicts the impact change of each input variable on the NPV and their sensitivity range. The analysis results show that in terms of NPV, the project's profitability is primarily most sensitive to the electricity sales price and production. The most impactful input variables are the discount rate, the initial investment, and the biomass cost (the least affecting). Electricity sales price and production significantly impact NPV compared to the remaining input variables. In fact, from the five input variables tested, electricity sales price and production may greatly compromise the project's economic viability in an unfavorable and stressful scenario by predicting a negative NPV up to €–6.6M and €–5.1M, respectively.

On the other hand, it may also considerably increase the project's current calculated revenue from €2.4M to a maximum of €11.2M and €10.0M. The electricity sales price is somewhat uncertain given its dependence on energy market price fluctuations and subsidies, which rely on political decisions. Thus, this is an utmost important factor to consider for the project's economic viability. As for the electricity production, it determines the annual output of the power plant, which heavily impacts the annual revenues, relying upon a stable annual production to avoid investment loss. The discount rate and initial investment share almost an identical impact on the NPV. Both are highly impactful, as the selected discount rate will influence the NPV discounted cash flow calculation, and the initial investment will cause a tremendous initial cash outflow. Biomass cost is the least consequent of the set. Even in a worsening price scenario, it still achieves an NPV revenue within the €2.1–2.6M range, reasonably close to the calculated deterministic value (€2.4M). Therefore, it can be verified that, unlike the remaining input variables, biomass cost is the only which does not compromise the NPV performance to a negative range scenario.

Overall, despite the project's viability and affordable risk provided by the economic model calculations, there is also the attractiveness of the investment to consider. The calculated IRR and PBP failed to meet the typical benchmarks of a 10% rate and less than 10 years of return period mentioned by the World Bank Group [11]. Also, the sensitivity analysis shows that from all the critical variables tested, the electricity sales price is the one that carries the most substantial impact on the NPV. The project's economic performance is highly dependent on revenues from electricity sales regulated by uncertain tariffs and reimbursements. Moreover, additional uncertainties related to technoeconomic calculation analysis must be considered. Project

Impact Change on NPV

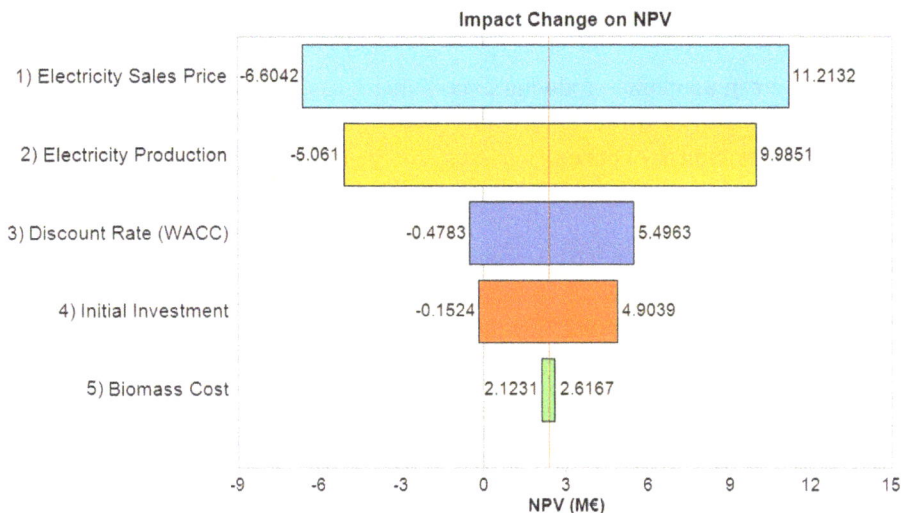

Figure 5.3: Sensitivity range to input variables for NPV.

cash flows are based on assertions taken from the literature and project evaluation reports once actual cash flows are estimated within a range of possibilities rather than certain single numbers. All these assumptions will add uncertainty to the already predicted risk share, leading investors to demand higher financial assurances before accepting the venture. Thus, it is important to look beyond the numbers and consider all features that might turn the project's viability the other way round.

5.5 Technoeconomic analysis of biomass gasification for green ammonia production

The promising nature of green ammonia and the need to contribute to its technological development have turned green ammonia research into one of the hottest topics for future energy applications [29] Furthermore, Portugal presents favorable and unique conditions for hydrogen and green ammonia production, owing to its strategic geographic location, high penetration of renewables, and very competitive renewable electricity production prices, placing Portugal as a strong contender in the European hydrogen economy [30].

Here, one extends the previously developed economic model to deliver a small-scale green ammonia production analysis through biomass gasification in mainland Portugal. The investment analysis is performed by employing a combined economic model coupling the NPV, IRR, modified internal rate of return (MIRR), PBP, and discounted payback period (DPBP).

5.6 Technoeconomic assessment

A 1 MW of green ammonia production via biomass gasification power plant is proposed for deployment, within a pulp and paper mill, in southeastern Portugal. The pulp and paper industry is one of the most energy-intensive sectors of the EU, facing increased energy prices and ever-demanding competition for raw materials, highlighting the need for developability and seeking additional revenue streams for this manufacturing industry [31]. Moreover, pulp and paper mills are par excellence hosts for these biomass systems by possessing existing infrastructures for the biomass residues supply and handling, fully exploiting these by-products, increasing the operations efficiency and profitability [32]. This study was built in line with the Portuguese National Hydrogen Strategy (EN-H$_2$) [30], current literature, and technical reports related to biomass-to-ammonia power plants [11, 31–34].

Located on the edge of the Atlantic Ocean, Sines hosts the main port on the Ibero-Atlantic front, the Port of Sines. Its strategic location integrates the required set of infrastructures for ammonia's reception and storage, transportation network (both maritime and land freight), and connection to the natural gas distribution network (Figure 5.4). The port's liquified natural gas terminal welcomes between 40,000 and 216,000 m^3 of natural gas (~59 ships/year) for an unload rate of 10,000 m^3/h. Storage capacity sets around 2569 GWh (total tank capacity of 390,000 m^3) and ship loading rate around 1500 m^3/h. Also, tank truck filling bays allow for rates around 585 m^3/h, setting land freight ready for dispatch to national regions not supplied by the pipeline network and across Europe. The Portuguese National Hydrogen Strategy (EN-H$_2$) establishes a strategic partnership in which ship route exports are first expected to occur to Northern Europe (mainly to the Netherlands) and then to other countries and regions. Conveniently, the already existing natural gas pipeline network (1375 km of main gas pipeline and high-pressure branches with an output capacity of 666 GWh/day, equivalent to a flow rate of 2330 km^3/h) can be adapted to transport liquid ammonia with minimal changes once ammonia requires far lower pressures than natural gas while having good compatibility with iron and steel materials that build the current infrastructures [30].

Besides these promising conditions, the considered area is also set close to an industrial park with current and future major hydrogen consumers (e.g., refineries and chemical industries) and benefits from ample construction land availability. Furthermore, the municipality of Sines lies within an important agricultural region, the Alentejo (area of about 3×10^6 ha), which is a supply market for ammonia for fertilization purposes. All these advantageous features have already earned Sines the commitment to the first green hydrogen production pilot plant (up until 2030) entrusted by the Portuguese Government [30]. Thus, in hand with this framework, the power plant is proposed for deployment in the Sine's industrial and port hub, facilitating the distribution of the plant's ammonia production to the local consumer industry, agricultural cooperatives, and for port dispatch.

5.7.1 Modified internal rate of return

As previously mentioned, the IRR assumes that the project's net cash inflows are reinvested at a rate as same as the IRR. In contrast, multiple IRR values may prevail for the same project, particularly if cash flows are irregular. To circumvent this, the MIRR is applied, which is defined as the rate of return leading the project's present value of terminal inflows to equal the outflows [47]:

$$\text{MIRR} = \sqrt[N]{\frac{\text{Future value of positive cash flows}}{-\text{Present value of negative cash flows}}} - 1 \tag{5.5}$$

Thus, the MIRR can be expressed as follows:

$$\text{MIRR} = \sqrt[N]{\frac{\sum_{n=1}^{N} C_n^+ (1 + k_{rr})^{N-n}}{-\sum_{n=0}^{N} \frac{C_n^-}{(1 + k_{fr})^n}}} - 1 \tag{5.6}$$

where C_n^+ and C_n^- represent the net cash inflow and outflow at time n, respectively, N is the total number of years, k_{rr} is the reinvestment rate, and k_{fr} is the finance rate. During the calculations, the k_{rr} and k_{fr} are assumed the same as the discount rate i.

5.7.2 Discounted payback period

The DPBP formulates less straightforwardly as PBP, granting a more accurate prediction as follows [45]:

$$\text{DPBP} = \frac{\ln\left(\frac{1}{1 - \frac{C_0 \times i}{C_t}}\right)}{\ln(1 + i)} \tag{5.7}$$

where C_0 is the initial investment, C_t is the annual net cash flow in the tth year, and i is the discount rate.

5.8 Biomass-to-ammonia economic model results and discussion

Figure 5.5 shows the economic model results for the proposed 1 MW green ammonia production via biomass gasification power plant. The current depiction shows the NPV profile variation as a function of the discount rate, the IRR at the moment that NPV equals zero, the MIRR, and the cumulative cash flow and cumulative NPV along the plant's lifetime delivering both the PBP and DPBP at the year a positive stage is reached. According to previous assumptions, the venture criterion of acceptance relies on a positive NPV, an IRR and MIRR higher than the discount rate, and a PBP and DPBP less than the power plant's lifetime. Under current market conditions, the investment project predicts a positive NPV of €3714k, an IRR of 24.32%, MIRR of 14.99% (both higher than the discount rate of 10%), a PBP of 4.6 years, and DPBP of 5.8 years (both less than the 20-year lifetime) to redeem the initial capital investments.

NPV Profile, IRR, MIRR, PBP and DPBP
(1 MW biomass gasification to ammonia power plant)

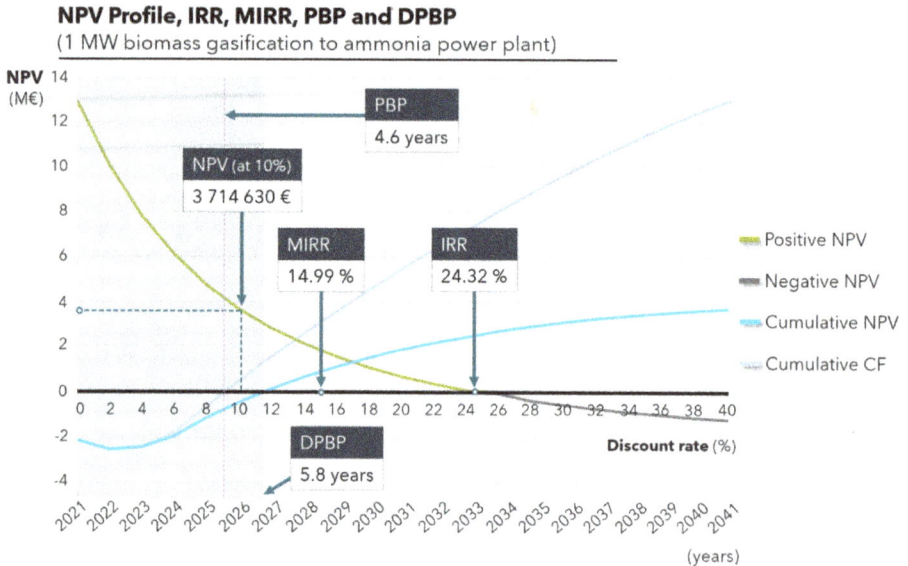

Figure 5.5: Economic model results containing the NPV profile, IRR, MIRR, PBP, and DPBP for the 1 MW ammonia production via biomass gasification power plant (CF: cash flow).

In general, the economic model outlays a positive scenario for this investment project, with all financial indicators satisfying the criterion for acceptance. However, as expected, the financial indicators IRR and MIRR, PBP, and DPBP deliver contrasting results, with the IRR and MIRR differing considerably (nearly in a 10% range), whereas the PBP and DPBP only differ in 1.2 years. In this work, the standard IRR tends to overestimate the expected profitability of the project by misinterpreting the transition between negative and positive cash flows (especially when shifting from the first to the following years), hence delivering multiple numerical solutions while assuming that the positive cash flows are reinvested at a much higher rate than the capital cost of the investment. Conversely, the MIRR considers a reinvestment rate equal to the applied discount rate accounting for a single solution regardless of the cash flow plan (positive or negative), which leads to a more realistic approach for this sort of investment. As for the payback methods, the DPBP considers the cumulative discounted net cash flows, meaning that earlier cash flows are worth more than later ones, thus acknowledging the time value of money. This cash flow depreciation confers DPBP a less favorable scenario than the PBP, yet it allows for a more accurate payback prediction.

As an attempt to bring these economic results into comparison, Table 5.5 provides some financial indicators gathered from other biomass-to-ammonia studies. The literature may be short on a broad technoeconomic analysis of biomass gasification systems used to produce green ammonia. Still, it is even shorter concerning applying combined NPV, IRR, MIRR, PBP, and DPBP economic models. To our knowledge, no MIRR and DPBP results were found concerning biomass-to-ammonia systems. In general, the financial indicators address similar trends to the ones drawn from the literature, specifically if the upgraded methods of MIRR and DPBP are weighted in, to which the 14.99% rate and 5.8 years consistently lie within the trends attained from the literature. In opposition, the NPV method varies greatly from project to project once these are highly erratic when differing investment amounts are applied. Concerning decision-making, cost estimates broadly return feasible conditions for biomass-to-ammonia power plants, despite these being highly reliant upon high capital costs, ammonia sales revenues, and production capacity [31, 40].

Table 5.5: Financial indicators for biomass-to-ammonia technology power plant.

Biomass-to-ammonia power plant NH$_3$ production capacity	NPV	IRR	PBP	Year	References
35,755 tons/year of NH$_3$	€0.15M	13.2%	20 years	2017	[48]
50,000 tons/year of NH$_3$	–	–	6–7 years	2020	[33]
228,000 tons/year of NH$_3$	–	10–20%[a]	–	2014	[31]
268,818 tons/year of NH$_3$	–	10%	–	2008	[49]
414,000 tons/year of NH$_3$	–	20%[b]	–	2014	[40]

[a]For an NH$_3$ selling price range of €581–882/ton.
[b]Target rate calculated based on 2000–2009 average NH$_3$ prices.

5.9 Monte Carlo sensitivity analysis

Figure 5.6 presents the sensitivity analysis range to the selected input variables. For simplification purposes, the impact is only considered for the NPV, as investment loss is more likely to occur through this financial indicator. The curves with the steepest slopes have the greatest sensitivity to variational change. Thus, ammonia production and sales price affect NPV performance the most, from a minimum of €2423k and €2611k to a maximum of €5008k and €4827k, respectively. These two variables are highly impactful as a stable ammonia production output is pivotal for profiting and thus avoids investment loss.

In contrast, ammonia sales prices can be somewhat uncertain as these rely heavily on market price fluctuations and subsidies. After that, the discount rate and biomass cost show a lower impact change on the NPV. However, as previously mentioned, these two variables can be impactful cost drivers as NPV is highly reliant on the discount rate, and biomass costs carry noticeable cash flows to maintain a stable annual production avoiding feed disruption. As a result, both variables share identical impact changes driving the NPV from a minimum of €3391k to a maximum of €4044k. On the other hand, O&M costs and initial investment are the least impactful, leading to a worsening NPV scenario of €3465k and a favorable NPV of €3962k.

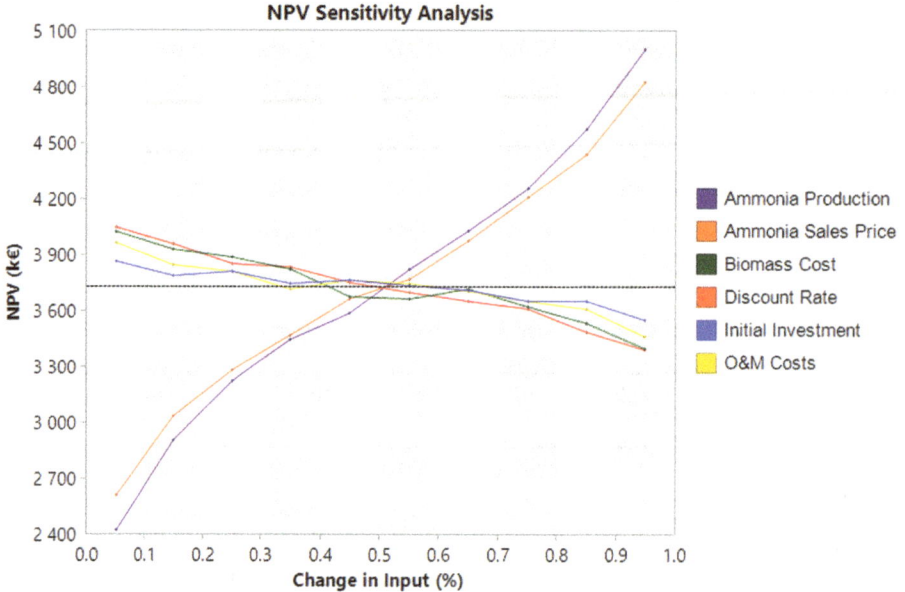

Figure 5.6: Sensitivity analysis range to input variables for NPV.

Finally, the studied financial indicators determined a low risk of investment loss. Yet, investment loss was more likely to occur due to NPV failure, with ammonia sales price and ammonia production having a greater possible impact on its performance. These variables are highly impactful as the first relies heavily on ammonia market price fluctuations and subsidies. At the same time, the latter must provide a stable ammonia production output resulting in important revenues for the project's economic success.

Additional analyses conclude that biomass gasification is a cost completive and environmentally effective energy source within certain application scenarios, averting the need for breakthrough advances in catalyst processes, delivering feedstock cost benefits and favorable economies of scale. This vision was met by the presented small-scale scheme, allowing it to avoid complex logistics and increased feedstock supply requirements intrinsic to large-scale units.

References

[1] Ecoprog, "Biomass to Power e the World Market for Biomass Power Plants," 2016.
[2] REM, "Power generation from biomass booms worldwide," Jul. 21, 2011. https://www.renew ableenergymagazine.com/biomass/power-generation-from-biomass-booms-worldwide (accessed Aug. 05, 2022).
[3] J. Cardoso, V. Silva, and D. Eusébio, "Techno-economic analysis of a biomass gasification power plant dealing with forestry residues blends for electricity production in Portugal," *Journal of Cleaner Production*, vol. 212, pp. 741–753, Mar. 2019, doi: 10.1016/ J.JCLEPRO.2018.12.054.
[4] S. A. Sulaiman, M. Inayat, H. Basri, F. M. Guangul, and S. M. Atnaw, "Effect of blending ratio on temperature profile and syngas composition of woody biomass co-gasification," 2016. http://eprints.utp.edu.my/25343/ (accessed Aug. 05, 2022).
[5] M. Carneiro, *Avaliaçao Economica da Biomassa para a Produçao de Energia*, University of Minho, 2010.
[6] P. C. J. Freitas, "Avaliação de Projecto de Investimento CENTRAL TERMOELÉCTRICA A BIOMASSA FLORESTAL (CTBF) Relatório de Avaliação," 2009.
[7] IRENA, "Biomass for Power Generation. Renewable Energy Technologies: Cost Analysis Series," 2012.
[8] NREL, "Biomass Gasification Technology Assessment. National Renewable Energy Laboratory.," 2012. https://www.nrel.gov/docs/fy13osti/57085.pdf (accessed Aug. 05, 2022).
[9] NuGen Engineering Ltd, "'Stewards of Our Traditional Lands' Kelly Lake, British Columbia Feasibility Study for a 10 MW Biomass Fired Power Plant," 2010.
[10] Y. Seo, H. S. Han, E. M. T. Bilek, J. Choi, D. Cha, and J. Lee, "Economic analysis of a small-sized combined heat and power plant using forest biomass in the Republic of Korea," vol. 13, no. 3, pp. 116–125, Jul. 2017, http://dx.doi.org/10.1080/21580103.2017. 1350209, doi: 10.1080/21580103.2017.1350209.
[11] World Bank, "International Finance Corporation. World Bank Group. Converting Biomass to Energy. A Guide for Developers and Investors.," 2017.
[12] Assembleia da República, "Relatorio Grupo de Trabalho da Biomassa. Comissao de Agricultura e Mar.," 2013.
[13] ERSE, "Proveitos Permitidos e Ajustamentos para 2018 das Empresas Reguladas do Setor Eletrico. Entidade Reguladora dos Serviços Energeticos, Lisboa.," 2017.
[14] EQTEC, "Waste & biomass gasification," 2018. https://eqtec.com/es/ (accessed Aug. 05, 2022).
[15] H. Viana, W. B. Cohen, D. Lopes, and J. Aranha, "Assessment of forest biomass for use as energy. GIS-based analysis of geographical availability and locations of wood-fired power plants in Portugal," *Applied Energy*, vol. 87, no. 8, pp. 2551–2560, Aug. 2010, doi: 10.1016/ J.APENERGY.2010.02.007.
[16] D. T. Pio, L. A. C. Tarelho, and M. A. A. Matos, "Characteristics of the gas produced during biomass direct gasification in an autothermal pilot-scale bubbling fluidized bed reactor," *Energy*, vol. 120, pp. 915–928, Feb. 2017, doi: 10.1016/J.ENERGY.2016.11.145.
[17] V. Silva, E. Monteiro, N. Couto, P. Brito, and A. Rouboa, "Analysis of syngas quality from Portuguese biomasses: An experimental and numerical study," *Energy and Fuels*, vol. 28, no. 9, pp. 5766–5777, Sep. 2014, doi: 10.1021/EF500570T.
[18] E. Searcy and P. Flynn, "The impact of biomass availability and processing cost on optimum size and processing technology selection," *Applied Biochemistry and Biotechnology*, vol. 154, no. 1–3, pp. 271–286, May. 2009, doi: 10.1007/S12010-008-8407-9.

[19] C. Franco, F. Pinto, I. Gulyurtlu, and I. Cabrita, "The study of reactions influencing the biomass steam gasification process☆," *Fuel*, vol. 82, no. 7, pp. 835–842, May. 2003, doi: 10.1016/S0016-2361(02)00313-7.

[20] A. Garg and M. P. Sharma, "Performance Evaluation of Gasifier Engine System Using Different Feed Stocks," 2013.

[21] P. Malatji, N. S. Mamphweli, and M. Meincken, "The technical pre-feasibility to use briquettes made from wood and agricultural waste for gasification in a downdraft gasifier for electricity generation," *Journal of Energy in Southern Africa*, vol. 22, pp. 2–7, Aug. 2011, [Online]. Available: http://www.scielo.org.za/scielo.php?script=sci_arttext&pid=S1021-447X2011000400001&nrm=iso.

[22] J. Cardoso, V. Silva, D. Eusébio, P. Brito, M. J. Hall, and L. Tarelho, "Comparative scaling analysis of two different sized pilot-scale fluidized bed reactors operating with biomass substrates," *Energy*, vol. 151, pp. 520–535, May. 2018, doi: 10.1016/J.ENERGY.2018.03.090.

[23] Bank of Portugal, "Economic projections, 25 July 2018," 2018.

[24] Statistics Portugal, "Consumer price Index," 2017.

[25] Y. Zhu, R. Zhai, Y. Yang, and M. Angel Reyes-Belmonte, "Techno-economic analysis of solar tower aided coal-fired power generation system," *Energies*, 2017, vol. 10, no. 9, p. 1392, Sep. 2017, doi: 10.3390/EN10091392.

[26] NETL, "Gasification Systems," 2010. https://netl.doe.gov/carbon-management/gasification (accessed Aug. 05, 2022).

[27] P. Lamers et al., "Techno-economic analysis of decentralized biomass processing depots," *Bioresource Technology*, vol. 194, pp. 205–213, Oct. 2015, doi: 10.1016/J.BIORTECH.2015.07.009.

[28] J. Sousa Cardoso, V. Silva, D. Eusébio, I. Lima Azevedo, and L. A. C. Tarelho, "Techno-economic analysis of forest biomass blends gasification for small-scale power production facilities in the Azores," *Fuel*, vol. 279, p. 118552, Nov. 2020, doi: 10.1016/J.FUEL.2020.118552.

[29] J. S. Cardoso, V. Silva, R. C. Rocha, M. J. Hall, M. Costa, and D. Eusébio, "Ammonia as an energy vector: Current and future prospects for low-carbon fuel applications in internal combustion engines," *Journal of Cleaner Production*, vol. 296, May. 2021, doi: 10.1016/J.JCLEPRO.2021.126562.

[30] Participa.pt, "EN-H2 Estratégia Nacional para o Hidrogénio," 2020. https://participa.pt/pt/consulta/en-h2-estrategia-nacional-para-o-hidrogenio (accessed Aug. 05, 2022).

[31] J. Andersson and J. Lundgren, "Techno-economic analysis of ammonia production via integrated biomass gasification," *Applied Energy*, vol. 130, pp. 484–490, Oct. 2014, doi: 10.1016/J.APENERGY.2014.02.029.

[32] M. Akbari, A. O. Oyedun, and A. Kumar, "Ammonia production from black liquor gasification and co-gasification with pulp and waste sludges: A techno-economic assessment," *Energy*, vol. 151, pp. 133–143, May. 2018, doi: 10.1016/J.ENERGY.2018.03.056.

[33] H. Zhang, L. Wang, J. van Herle, F. Maréchal, and U. Desideri, "Techno-economic comparison of green ammonia production processes," *Applied Energy*, vol. 259, pp. 114135, Feb. 2020, doi: 10.1016/J.APENERGY.2019.114135.

[34] A. Sánchez, M. Martín, and P. Vega, "Biomass based sustainable ammonia production: Digestion vs gasification," *ACS Sustainable Chemistry and Engineering*, vol. 7, no. 11, pp. 9995–10007, Jun. 2019, doi: 10.1021/ACSSUSCHEMENG.9B01158/SUPPL_FILE/SC9B01158_SI_001.PDF.

[35] L. J. R. Nunes, J. C. O. Matias, and J. P. S. Catalão, "Biomass waste co-firing with coal applied to the Sines Thermal Power Plant in Portugal," *Fuel*, vol. 132, pp. 153–157, Sep. 2014, doi: 10.1016/J.FUEL.2014.04.088.

[36] N. Couto, V. Silva, E. Monteiro, and A. Rouboa, "Exergy analysis of Portuguese municipal solid waste treatment via steam gasification," *Energy Conversion and Management*, vol. 134, pp. 235–246, 2017, doi: 10.1016/J.ENCONMAN.2016.12.040.

[37] V. Silva and A. Rouboa, "Optimizing the gasification operating conditions of forest residues by coupling a two-stage equilibrium model with a response surface methodology," *Fuel Processing Technology*, vol. 122, pp. 163–169, Jun. 2014, doi: 10.1016/J.FUPROC.2014.01.038.

[38] J. S. Cardoso, V. Silva, J. A. M. Chavando, D. Eusébio, M. J. Hall, and M. Costa, "Small-scale biomass gasification for green ammonia production in Portugal: A techno-economic study," *Energy & Fuels*, vol. 35, no. 17, pp. 13847–13862, Sep. 2021, doi: 10.1021/ACS.ENERGYFUELS.1C01928.

[39] J. Cardoso, V. Silva, D. Eusébio, P. Brito, and L. Tarelho, "Improved numerical approaches to predict hydrodynamics in a pilot-scale bubbling fluidized bed biomass reactor: A numerical study with experimental validation," *Energy Conversion and Management*, vol. 156, pp. 53–67, Jan. 2018, doi: 10.1016/j.enconman.2017.11.005.

[40] P. Gilbert, S. Alexander, P. Thornley, and J. Brammer, "Assessing economically viable carbon reductions for the production of ammonia from biomass gasification," *Journal of Cleaner Production*, vol. 64, pp. 581–589, Feb. 2014, doi: 10.1016/J.JCLEPRO.2013.09.011.

[41] P. Arora, A. F. A. Hoadley, S. M. Mahajani, and A. Ganesh, "Small-scale ammonia production from biomass: A techno-enviro-economic perspective," *Industrial & Engineering Chemistry Research*, vol. 55, no. 22, pp. 6422–6434, Jun. 2016, doi: 10.1021/ACS.IECR.5B04937/SUPPL_FILE/IE5B04937_SI_002.ZIP.

[42] B. Silva, "Biomassa duplica preço. Novas centrais podem falir antes de abrir portas," Feb. 08, 2019. https://www.dinheirovivo.pt/economia/biomassa-duplica-preco-novas-centrais-podem-falir-antes-de-abrir-portas-12782817.html (accessed Aug. 05, 2022).

[43] Banco de Portugal, "Projeções económicas," 2022. https://www.bportugal.pt/page/proje coes-economicas (accessed Aug. 05, 2022).

[44] D. Fioriti, S. Pintus, G. Lutzemberger, and D. Poli, "Economic multi-objective approach to design off-grid microgrids: A support for business decision making," *Renewable Energy*, vol. 159, pp. 693–704, Oct. 2020, doi: 10.1016/J.RENENE.2020.05.154.

[45] N. Indrawan, B. Simkins, A. Kumar, and R. L. Huhnke, "Economics of distributed power generation via gasification of biomass and municipal solid waste," *Energies*, vol. 13, no. 14, p. 3703, Jul. 2020, doi: 10.3390/EN13143703.

[46] H. Kierulff, "MIRR: A better measure," *Business Horizons*, vol. 51, no. 4, pp. 321–329, Jul. 2008, doi: 10.1016/J.BUSHOR.2008.02.005.

[47] A. Nicita, G. Maggio, A. P. F. Andaloro, and G. Squadrito, "Green hydrogen as feedstock: Financial analysis of a photovoltaic-powered electrolysis plant," *International Journal of Hydrogen Energy*, vol. 45, no. 20, pp. 11395–11408, Apr. 2020, doi: 10.1016/J.IJHYDENE.2020.02.062.

[48] G. Editors et al., "Technical and economic analysis for production of bio-ammonia from African Palm Rachis," *Chemical Engineering Transactions*, vol. 57, 2017, doi: 10.3303/CET17571160.

[49] J. R. Bartels, *A Feasibility Study of Implementing an Ammonia Economy*, 2008, doi: 10.31274/ETD-180810-1374.

6 Ammonia, coal, and biomass co-firing in a pilot-scale fluidized bed reactor

Conventional stationary power stations rely heavily on fossil fuels, compromising the ambitious environmental goals proudly announced in major conventions worldwide. To this day, coal-fired power generation still supplies about 30% of the world's primary energy and 41% of global electricity generation [1]. The current challenges of reducing this coal dependence have been raising interest in cleaner fuel solutions.

Ammonia (NH_3) is one of the most important commodity chemicals, holding a mature storage and distribution worldwide network with increasing importance to pivotal sectors of the global society such as fertilizer production. As a carbon-free fuel, NH_3 carries great potential with promising applications in energy systems [2, 3]. Production-wise, NH_3 can be attained exclusively via renewable energy sources, commonly known as green NH_3, triggering exploration as a route to the transition of NH_3 as the next sustainable fuel solution for power generation [4]. Apart from this, NH_3 operates as a prime hydrogen (H_2) carrier, standing as a highly efficient energy vector with high energy density, and an established and flexible infrastructure capable of mitigating key drawbacks of H_2 [5, 6]. Thus, the use of NH_3 brings major benefits in terms of deploying cost-effective H_2 storage solutions as well as backup production of power.

The co-firing of coal and NH_3 is an example of such promising alternative technologies, allowing the retrofitting of coal power facilities without major modifications while contributing to global decarbonization goals [7]. Therefore, coal–NH_3 co-firing is deemed a feasible and short-term solution, easily implemented within a reasonable time period, with a goal of effectively reducing CO_2 emissions. This allows existing coal-burning power plants to replace some of their fuel with carbon-free NH_3, delivering immediate greenhouse gas (GHG) emission reduction and thus extending the economic life of coal-fired assets in a carbon-constrained jurisdiction [8]. However, the use of NH_3 combustion for power generation still presents some limitations and research gaps that need to be filled, such as (1) low burning velocities compared with conventional fuels, (2) longer ignition delay times, (3) superficial knowledge on NH_3 blends flame structure and its characteristics, (4) prevalence of high nitrogen oxide (NO_x) emissions, and (5) slipped unburned NH_3 issues [2, 6, 9].

To the best of our knowledge, no biomass–NH_3 co-firing studies can be found. However, the Japanese IHI Corporation currently plans to conduct a technical evaluation and feasibility study for the ammonia–biomass co-firing in a coal-fired thermal power plant in Malaysia and Southeast Asia [10]. As for the co-firing of coal and NH_3, its application in coal-fired thermal power plants has already been attempted through various approaches. Furthermore, experimental and numerical studies concerning the detailed flame characteristics of NH_3–coal blends are still sparse [11]. For instance, Chugoku Electric Power Company conducted the first NH_3–coal co-firing in 2017 in a 156-MW power plant using a fuel mix ranging between 0.6% and 0.8% NH_3. The

https://doi.org/10.1515/9783110758214-006

findings indicated that the addition of NH_3 did not affect the plant's power efficiency and reduce carbon dioxide emissions [12]. Indeed, Japan is leading the increased interest around this subject matter. More recently, the IHI Corporation demonstrated the co-firing of NH_3 and coal in a fuel mix composed of 20% NH_3 [13]. Results showed that with NH_3 co-firing, NO_x emission and unburned carbon (UC) in fly ash could reach the same level as in coal-only combustion if proper NH_3 injection is employed.

To the best of our knowledge, no numerical studies can be found in the literature regarding coal–NH_3 co-firing performed in pilot-scale fluidized bed reactors, and no studies can be found concerning the biomass inclusion within this fuel mix. Therefore, this chapter aims to bridge the current research gap concerning this topic, by seeking to contribute to the various strategies attempting to deliver decreased carbon emissions for coal-based facilities and provide supplementary resource options to coal-based power plants through NH_3 co-firing strategies. To do so, this chapter commits to:

-   assess the co-firing of NH_3 ratios and determine its effect on the combustion process;
-   determine the NH_3 co-firing ratio effect on coal combustion emissions reduction; and
-   finally, study the co-firing of coal–biomass–NH_3 so to assess its effect on carbon and NO emissions reduction from coal-fired power systems.

Furthermore, fluidized bed gasification is considered to be one of the most advanced methods for thermochemical conversion of various biomass fuels to energy offering economic and environmental benefits. For that reason and given the disruptive character of NH_3's use in fluidized bed reactors, this chapter delves into the co-combustion of coal, biomass, and NH_3 instead of its gasification. Such will be addressed in future publications.

Therefore, a 2D Eulerian–Lagrangian numerical model set within the ANSYS Fluent framework is used to describe the co-firing of coal, biomass, and NH_3 in a pilot-scale bubbling fluidized bed reactor. The numerical model is validated against experimental data on coal and biomass combustion to assess the accuracy of the model.

6.1 Experimental conditions

6.1.1 Combustion facility description and experimental conditions

Experiments were conducted in a thermally insulated pilot-scale bubbling fluidized bed reactor measuring 0.25 m wide by 2.3 m tall. Figure 6.1 shows a detailed schematic of the combusting facility. A computer-based control and data gathering system was used to monitor operational parameters such as feedstock input rates, airflow rates, temperature, and flue gas composition. Quartz sand was used and bed material with a density of 2650 kg/m^3 and particle size around 500 μm. Bituminous coal and forest residues were fed into the reactor at a rate of 17–50 g/min from 0.35 m above the distribution plate. Bituminous coal held a density of 1346 kg/m^3 and was sieved to a particle size of 500–4000 μm, while forest residues had a density of 478 kg/m^3 and particle size ranging from 1000 to 4000 μm. Table 6.1 shows the properties of feedstocks. Regarding the airflow, the first stream was heated and supplied from the bed's bottom, keeping a fluidization speed between 0.28 and 0.32 m/s depending on the operating condition. At the same time, the second stream was fed about 0.30 m above the bed surface. Thus, both streams provided a total airflow of 250 NL/min. Furthermore, eight water-cooled probes arranged along the bed maintained the bed temperature within the desired range. The excess air and air staging were established by adjusting the coal feed and secondary airflow rates. Air staging was used, with the primary air at the bottom accounting for 100%, 80%, and 60% of the total combustion air. Experiments were carried out under steady-state conditions with three surplus air levels (10%, 25%, and 50%) and three operating temperatures (1023, 1098, and 1173 K). Gas samples were collected and passed through a range of detectors, including paramagnetic (O_2), nondispersive infrared (CO_2, CO, N_2O, SO_2), chemiluminescence (NO), and flame ionization (hydrocarbons). Further details concerning the experimental setup and combustion facility can be found in [14, 15].

Table 6.1: Chemical composition of coal [14, 15].

	Bituminous coal	Forest residues
Proximate analysis (wt.%)		
Moisture	4.4	9.0
Ash	13.8	2.0
Volatile matter	28.5	70.0
Fixed carbon	53.3	19.0
Ultimate analysis (wt.%) (as given)		
Ash	13.8	2.5
C	65.97	49.8
H	4.27	6.7
N	1.67	3.0
S	0.88	n.d.
O (by difference)	13.41	38.0

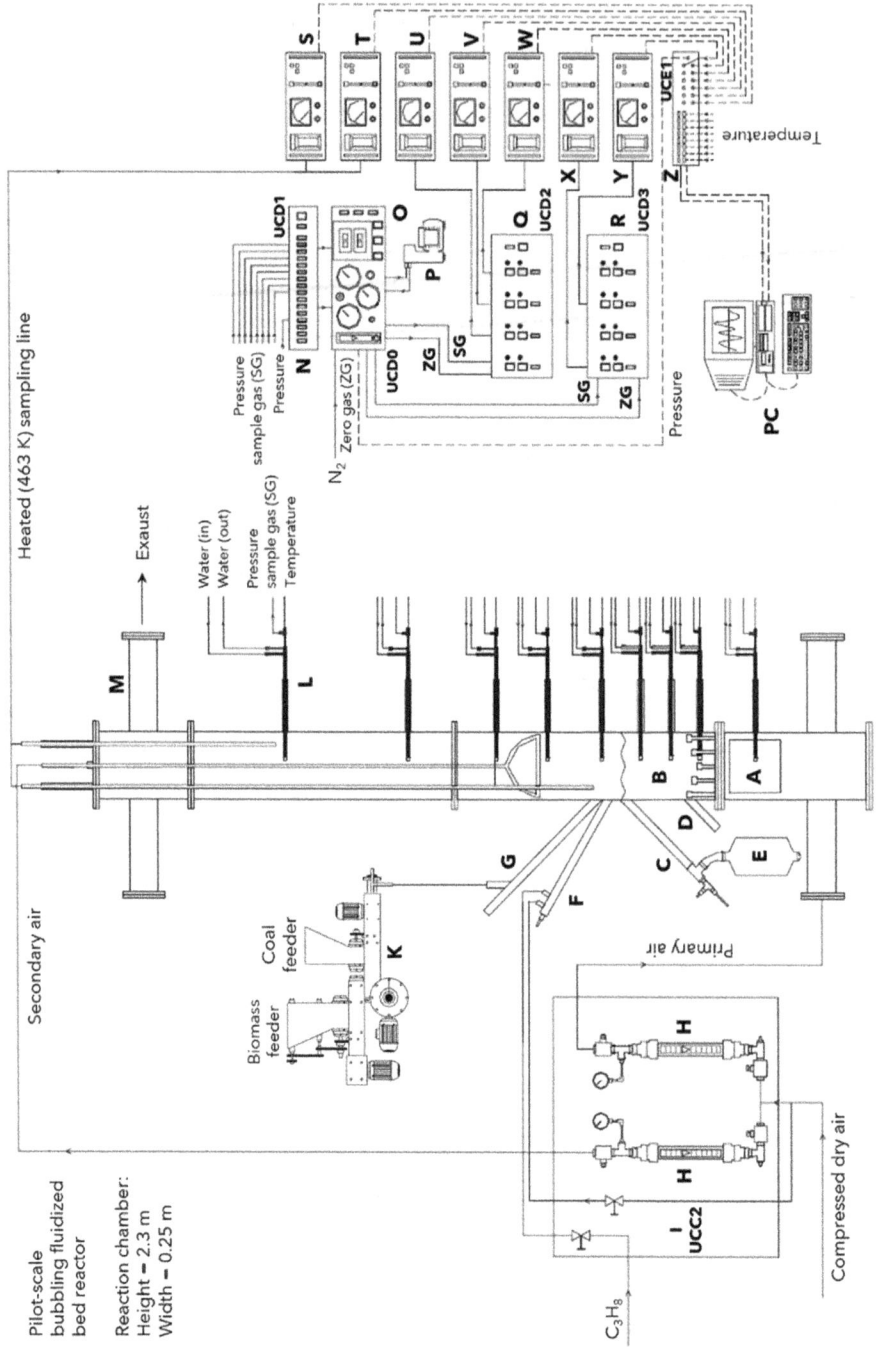

Figure 6.1: Detailed schematic of the experimental combustion unit: Dashed line, electric circuit; continuous line, pneumatic circuit; A, primary air heating system; B, sand bed;

6.2 Ammonia–coal–biomass co-firing mathematical model

The 2D numerical simulation was performed using Fluent, in which the gas phase was treated under the Eulerian framework while the solid phase was individually tracked under the Lagrangian method. The simplified 2D computational geometry domain was designed with a width of 0.25 m and height of 2.3 m and conceived to simulate the actual experimental conditions as closely as possible, as described in Figure 6.1. The primary atmospheric air was delivered at the bottom of the vessel, while both the secondary air and NH_3 inlets were set at the reactor's left sidewall. NH_3 co-firing ratio simulations were performed from 0% to 80% (by mass), where 0% portraits pure coal firing. The resulting flue gas is removed through the outlet placed at the top right corner of the geometry. The main interactions between the gas and solid phases were modeled by treating the mass exchange, momentum, and energy. The governing equations are briefly described in the following sections.

6.2.1 Continuous gas phase

The continuity equation for the gas-phase mass balance is given as follows [16]:

$$\frac{\partial}{\partial t}\left(\alpha_g \rho_g\right) + \nabla \cdot \left(\alpha_g \rho_g \vec{v}_g\right) = \delta \dot{m}_s \tag{6.1}$$

where v is the instantaneous velocity of gas phase, ρ is the density, α is the volume fraction, and $\delta \dot{m}_s$ is the gas mass production rate per unit volume from particle-gas chemistry (the subscript g refers to the gas phase).

The gas-phase momentum equation is depicted in eq. (6.2), where p_g is the mean flow gas thermodynamic pressure, F is the rate of momentum exchange per unit volume between the gas and solid phases, g is the gravitational acceleration, and τ_g is the gas phase stress tensor [16]:

$$\frac{\partial}{\partial t}\left(\alpha_g \rho_g \vec{v}_g\right) + \nabla \cdot \left(\alpha_g \rho_g \vec{v}_g \vec{v}_g\right) = -\alpha_g \cdot \nabla p_g + F + \alpha_g \rho_g g + \nabla \cdot \left(\alpha_g \tau_g\right) \tag{6.2}$$

Figure 6.1 (continued)
C, bed solid level control; D, bed solid discharge; E, bed solid discharge silo; F, propane burner system; G, coal addition port; H, airflow meter; I, control and command unit (UCC2); K, coal and biomass feeder; L, gas sampling probe; M, exhaust duct; N, O, Q, R, command and gas distribution units; P, gas sampling pump; PC, computer data acquisition and control system; S, T, U, V, W, X, Y, automatic online gas analyzers for determination of O_2, CO_2, CO, N_2O, NO, SO_2, HC; Z, electronic command unit; UCD0, UCD1, UCD2, UCD3, electropneumatic command and gas distribution units; UCE1, electronic command unit.

The energy conservation equation for the gas phase is as follows [16]:

$$\frac{\partial}{\partial t}\left(\alpha_g \rho_g h_g\right) + \nabla \cdot \left(\alpha_g \rho_g \vec{v}_g h_g\right) = \alpha_g \left(\frac{\partial p}{\partial t} + \vec{v}_g \cdot \nabla p_g\right) + \phi - \nabla \cdot \left(\alpha_g q\right) + \dot{Q} + S_h + \dot{q}_D \quad (6.3)$$

where α_g is the gas volume fraction, ρ_g is the gas density, h_g is the specific gas enthalpy, $\vec{v}_g$ is the gas velocity, ϕ is the viscous dissipation, $\dot{Q}$ represents the energy source per unit volume, q represents the heat flux, S_h is the conservative energy exchange from solid to gas phase, and $\dot{q}_D$ is the enthalpy diffusion term. The gas heat flux is given by

$$q = -\lambda_g \nabla T_g \quad (6.4)$$

where λ_g is the gas thermal conductivity consisting of molecular conductivity and eddy conductivity from Reynolds stress. The eddy conductivity correlates with the turbulent Prandtl number through $Pr_t = C_p \mu_t / \lambda_t$, where C_p is the specific heat capacity, and μ_t is the turbulent viscosity (the standard Ansys Fluent Prandtl number of 0.85 is employed) [17]. The mixture enthalpy is associated with the species enthalpies:

$$h_g = \sum_{i=1}^{N_i} Y_{g,i} h_i \quad (6.5)$$

where N_i is the sum of all gas species. The species enthalpy relies on the gas temperature by

$$h_i = \int_{T_0}^{T_f} C_{p,i} dT + \Delta h_{f,i} \quad (6.6)$$

where $\Delta h_{f,i}$ is the heat of formation at a reference temperature T_0, $C_{p,i}$ is the specific heat capacity at constant pressure for species i. The pressure is given by eq. (6.7), where R is the universal gas constant, T_g is the temperature of the gas phase, $Y_{g,i}$ is the mass fraction, and Mw_i is the molecular weight of each species:

$$p = \rho_g R T_g \sum_{i}^{N_i} \frac{Y_{g,i}}{Mw_i} \quad (6.7)$$

The species transport equation for the gas phase is expressed by the following expression [16]:

$$\frac{\partial}{\partial t}\left(\alpha_g \rho_g Y_{g,i}\right) + \nabla \cdot \left(\alpha_g \rho_g Y_{g,i} \vec{v}_g\right) = \nabla \cdot \left(\rho_g D \alpha_g \nabla Y_{g,i}\right) + \delta \dot{m}_{i,chem} \quad (6.8)$$

where, $\delta \dot{m}_{i,chem}$ is the gas-phase net production rate through chemical reactions, and D refers to the turbulent mass diffusivity related to the viscous Schmidt number

(Sc) correlation, $\mu/\rho_g D = \mathrm{Sc}$, where μ is the dynamic viscosity (the standard of 0.7 is assumed for the Schmidt number) [17]. The considered gas-phase chemical species were O_2, CO_2, H_2O, H_2, CO, NO, NH_3, and N_2.

6.2.2 Discrete solid phase

The Lagrangian approach assumes that the solid fuel particles, consisting of a mixture of volatile matter, char, and ash, can be portrayed as spherical particles with a given size distribution. The discrete-phase species here considered are coal and biomass (tracked in the solid phase), and NH_3 (tracked in the liquid phase). The governing energy equation for particles can be equated as follows [18]:

$$m_i c_i \frac{dT_i}{dt} = h_i A_{pi}\left(T_g - T_i\right) + \frac{\varepsilon_i A_{pi}}{4}\left(G - 4\sigma T_i^4\right) - H_{\text{evap}} dm_{\text{vapor}} - H_{O_2/CO_2/H_2O} dm_{C - O_2/CO_2/H_2O}$$

(6.9)

where m_i is the mass of species i, c_i is the specific heat, T_g is the gas temperature, T_i is the temperature of species i, h_i is the interphase thermal transfer coefficient, ε_i is particle emissivity, A_{pi} is the external surface area, G is the incident radiation, σ is the Stefan–Boltzmann constant, and H is the heat of reaction to evaporate water or one of the three heterogeneous char species. The incident radiation (G) value is initially set to zero, and the discrete-ordinates (DO) radiation model was employed to solve the transport equation. The DO model is the most comprehensive radiation model and is commonly used to solve co-combustion CFD applications [1]. The absorption coefficient was calculated based on the weighted sum of gray gases model [17].

The general governing equations depicting the momentum are [18]

$$\frac{dv_i}{dt} = f_{D,i} + g\left(1 - \frac{\rho_g}{\rho_i}\right)$$

(6.10)

$$f_{D,i} = \frac{3\mu_g C_D \mathrm{Re}_p}{4\rho_i d_i^2}\left(v_g - v_i\right)$$

(6.11)

$$C_D = \begin{cases} \frac{24}{\mathrm{Re}p}\left(1 + \frac{1}{6}\mathrm{Re}_p^{\frac{2}{3}}\right) & \mathrm{Re}_p \le 1000 \\ 0.424 & \mathrm{Re}_p > 1000 \end{cases}$$

(6.12)

$$\mathrm{Re}_p = \frac{\rho_g d_i |v_g - v_i|}{\mu_g}$$

(6.13)

where $f_{D,i}$ is the drag per unit mass, C_D is the drag force coefficient, Re_p is the Reynolds number, g is the gravitational acceleration, ρ_g is the gas density, ρ_i is the density of particle i, μ_g is the gas viscosity, d_i is the diameter of particle i, v_g is the velocity of the gas, and v_i is the velocity of particle i. Given the aims of the study, it is assumed that there are no particle–particle interactions and that the interactions of the particles on the wall can be neglected.

The solid particle's mass governing equation can be solved through [18]

$$\frac{dm_i}{dt} = \frac{dm_{vapor}}{dt} + \frac{dm_{devol}}{dt} + \frac{dm_{C-O_2}}{dt} + \frac{dm_{C-CO_2}}{dt} + \frac{dm_{C-H_2O}}{dt} \tag{6.14}$$

where dm_{vapor}, dm_{devol}, dm_{C-O_2}, dm_{C-CO_2}, and dm_{C-H_2O} represent the mass change in particle i from water vapor loss, devolatilization, and char reactions, respectively.

6.2.3 Turbulence model

The k–ε realizable model was chosen for the turbulence model as it adds enhanced prediction to the already robust and reasonably accurate standard k–ε model. k is the turbulence kinetic energy and ε is the dissipation rate and these are determined by the following transport equations [17]:

$$\frac{\partial}{\partial t}(\rho k) + \frac{\partial}{\partial x_j}\left(\rho k u_j\right) = \frac{\partial}{\partial x_j}\left[\left(\mu + \frac{\mu_t}{\sigma_k}\right)\frac{\partial k}{\partial x_j}\right] + G_k + G_b - \rho\varepsilon - Y_M + S_k \tag{6.15}$$

$$\frac{\partial}{\partial t}(\rho\varepsilon) + \frac{\partial}{\partial x_j}\left(\rho\varepsilon u_j\right) = \frac{\partial}{\partial x_j}\left[\left(\mu + \frac{\mu_t}{\sigma_\varepsilon}\right)\frac{\partial\varepsilon}{\partial x_j}\right] + \rho C_1 S_\varepsilon - \rho C_2\frac{\varepsilon^2}{k + \sqrt{v\varepsilon}} + C_{1\varepsilon}\frac{\varepsilon}{k}C_{3\varepsilon}G_b + S_\varepsilon \tag{6.16}$$

where $C_1 = \max\left[0.43, \frac{\eta}{\eta+5}\right]$, $\eta = S\frac{k}{\varepsilon}$, and $S = \sqrt{2S_{ij}S_{ij}}$. G_k represents the generation of turbulence kinetic energy due to the mean velocity gradients, G_b is the generation of turbulence kinetic energy due to buoyancy, and Y_M is the contribution of the fluctuating dilatation in compressible turbulence to the overall dissipation rate. $\sigma_k = 1.0$ and $\sigma_\varepsilon = 1.2$ are the turbulent Prandtl numbers for k and ε, respectively; S_k and S_ε are user-defined source terms. $C_{1\varepsilon} = 1.44$, $C_2 = 1.9$, and $C_{3\varepsilon} = 0$ are constants suggested by Launder and Spalding [19]. μ_t is the gas-phase turbulent viscosity, which is computed as a function of k and ε:

$$\mu_t = \rho C_\mu\frac{k^2}{\in} \tag{6.17}$$

where C_μ is a constant set as 0.09 (the standard Ansys Fluent value).

6.2.4 Chemical reaction model

Table 6.2 summarizes the chemical reactions and corresponding rate coefficients used in the simulations. For coal and biomass devolatilization, the two competing rate models were employed, and char combustion was dealt with the multisurface reaction model [17]. For modeling homogeneous gas-phase reactions, the kinetic and the turbulent mixing rate effects of the gas phase are considered [20]. The chaotic fluctuations of solid particles condition the gaseous species leading to velocity and pressure fluctuations. To address this turbulent flow, the finite-rate/eddy-dissipation model was used to calculate reaction rates. The chemical reaction rate coefficients are based on the Arrhenius equation. The Arrhenius rates and the kinetic parameters for the reactions were drawn from literature [1, 21]. Heterogeneous reactions are influenced by many factors such as reactant diffusion, breakup of the char, and the interaction of reactions and turbulent flow. The interactions of these char reactions with gas species such as O_2 and H_2O are complex processes involving balancing the rate of mass diffusion of the oxidizing chemical species to the surface of fuel particles with the surface reaction of these species with the char. The overall reaction rate of a char particle is determined by the oxygen diffusion to the particle surface and the rates of surface reactions, which rely on the temperature and composition of the gaseous environment and on the particle size, porosity, and temperature. NO_x modeling was performed using a postprocessing method set within the ANSYS Fluent framework. Here, the reactions concerning NH_3, thermal NO_x, HCN formation, and NO_x reduction by char were solved.

Table 6.2: Chemical reaction model.

Reactions	Arrhenius reaction rates
Volatile combustion [22, 23]:	
$R_{1(coal)} = C_xH_yO_z + (x+y/4-z/2)O_2$ $\rightarrow xCO_2 + (y/2)H_2O$	$r_1 = 657{,}000\ T \exp\left(\frac{-8.02\times10^7}{T}\right)\left[C_xH_yO_z\right]^{0.5}[O_2]$
$R_{2(biomass)} = C_xH_yO_z + (x+y/4-z/2)O_2$ $\rightarrow xCO_2 + (y/2)H_2O$	$r_2 = 9.2\times10^8\ T \exp\left(\frac{-1.3576\times10^8}{T}\right)\left[C_xH_yO_z\right]^{0.5}[O_2]$
Homogeneous reactions[a]	
H_2 combustion [24]:	
$R_3 = H_2 + 0.5O_2 \leftrightarrow H_2O$	$r_2 = 5.69\times10^{11} \exp\left(\frac{-1.47\times10^8}{T}\right)[H_2][O_2]^{0.5}$
CO combustion [25]:	
$R_4 = CO + 0.5O_2 \rightarrow CO_2$	$r_3 = 1.93\times10^{13}\ T^{-2}\exp\left(\frac{-1.26\times10^8}{T}\right)[CO][O_2]^{0.5}$
Water-gas shift [24]:	
$R_5 = CO + H_2O \leftrightarrow CO_2 + H_2$	$r_4 = 2.75\times10^9 \exp\left(\frac{-8.36\times10^7}{T}\right)[CO][H_2O]$
NH_3 combustion [26]:	
$R_6 = NH_3 + O_2 \rightarrow NO + H_2O + 0.5H_2$	$r_5 = 350T^{7.65} \exp\left(\frac{-5.24\times10^8}{T}\right)[NH_3][O_2]$

Table 6.2 (continued)

Reactions	Arrhenius reaction rates
$R_7 = NH_3 + NO \rightarrow N_2 + H_2O + 0.5H_2$	$r_6 = 4.24 \times 10^5 T^{5.3} \exp\left(\frac{-3.5 \times 10^8}{T}\right)[NH_3][NO]$
NH$_3$ pyrolysis [27]:	
$R_8 = NH_3 \rightarrow 0.5N_2 + 1.5H_2$	$r_7 = 0.185T^{1.25} \exp\left(\frac{-6.9 \times 10^7}{T}\right)[NH_3]$
Heterogeneous reactions[b]	
Char combustion [28]:	
$R_9 = C(s) + 0.5O_2 \rightarrow CO$	$r_8 = 0.052 \exp\left(\frac{-6.1 \times 10^7}{T}\right)$
Boudouard reaction [29]:	
$R_{10} = C(s) + CO_2 \rightarrow 2CO$	$r_9 = 0.0732 \exp\left(\frac{-1.125 \times 10^8}{T}\right)$
Water-gas reaction [28]:	
$R_{11} = C(s) + H_2O \rightarrow CO + H_2$	$r_{10} = 0.0782 \exp\left(\frac{-1.15 \times 10^8}{T}\right)$
Eddy-dissipation rate	**Arrhenius equation**
$r_{eddy\ dissipation} = \alpha_{i,r} M_{w,i} A \rho \frac{\varepsilon}{k} \min$	$r_{Arrhenius} = A \exp\left(\frac{-E_a}{RT}\right)$
$\left(\min_R\left(\frac{Y_R}{\alpha_{R,r} M_{w,R}}\right), B\frac{\sum_p Y_p}{\sum_i^N \alpha_{i,r} M_{w,i}}\right)$	

Thermal NO$_x$ formation [17]	HCN route [17]	NO$_x$ reduction [17]
$N_2 + O \leftrightarrow NO + N$	$HCN + O_2 \rightarrow NO + 0.5H_2 + CO$	$C + NO \rightarrow 0.5N_2 + CO$
$N + O_2 \leftrightarrow NO + O$	$HCN + NO \rightarrow N_2 + 0.5H_2 + CO$	
$N + OH \leftrightarrow NO + H$		

[a]Arrhenius rate pre-exponential factor, A (s^{-1}); activation energy, E_a (J/kmol); rate of reaction, r (kg mol/m^3 s).
[b]A (kg/m^2 s Pa); E_a (J/kmol).

6.3 Results and discussion

6.3.1 Effect of NH$_3$ co-firing ratio on emission reduction

NH$_3$ injection has been pursued as a key approach for CO$_2$ emissions reduction from industrial combustion energy systems. This approach is validated by the decreasing trend of CO$_2$ emissions at the reactor's outlet with increases in the NH$_3$ co-firing ratio, as shown in Figure 6.2a. Since the composition of NH$_3$ is free of any carbon, its replacement in coal combustion directly leads to CO$_2$ emissions reduction by eliminating coal carbon as given by the combustion equation of coal volatiles ($C_xH_yO_z + O_2 \rightarrow CO_2 + H_2O$; see balanced reaction in Table 6.2). Further-more, NH_3 injection in the flue gas is known for inhibiting the CO oxidation to CO_2 ($CO + 0.5O_2 \rightarrow CO_2$) [30]. At a 20% co-firing ratio, CO_2 emissions decreased up to 26% compared to pure coal firing, meeting the 20% range addressed in the literature [8].

Figure 6.2b presents the normalized CO emissions. CO emissions tend to re-duce with both coal feed reduction and with increased NH_3 co-firing ratio. The water-gas shift reaction ($CO + H_2O \rightarrow CO_2 + H_2$) is also expected to play a signifi-cant role in decreasing CO emissions as does the NH_3 combustion (portrayed by the reactions $NH_3 + O_2 \rightarrow NO + H_2O + 0.5H_2$ and $NH_3 + NO \rightarrow N_2 + H_2O + 0.5H_2$); H_2O concentration broadly increases in the flue gas, due to the higher hydrogen con-tent of the overall feed, promoting the CO to react with H_2O and consequently leading to its abatement [8].

Figure 6.2c shows the normalized UC in fly ash with concentration levels de-creasing with the NH_3 ratio increase. The presence of UC in the ash results from in-complete coal combustion processes and is commonly associated with carbon loss and efficiency decrease. In this system, NH_3 injection is through a sidewall inlet. UC abatement through NH_3 co-firing in coal-fired systems has already been reported in the literature, where sidewall NH_3 injection showed enhanced UC reduction com-pared to burner or nozzle injection [7, 31]. Moreover, the present work was generally performed under air-rich conditions with an overall excess air ratio around 1.25. As a result, air-rich operating conditions are considered to lower UC concentrations [7].

As for the NO emissions at the reactor outlet (Figure 6.2d), NO concentration remains much the same as in pure coal firing until a 10% co-firing ratio is reached. The concentrations were normalized to 5% O_2 in the gases to correct for dilution effects [14]. A slight increase in the NO level is measured at a 20% ratio, which is commonly attributed to more intense combustion and an increase of fuel O_x from NH_3 injection. In other words, this indicates that fuel NO_x was produced in the flame by the oxidation of ammonia, or coal combustion was aggravated during this stage, resulting in the increase of NO emissions [1, 31]. When exceeding the 20% ratio up to 80%, NO emissions gradually decrease as a result of the $DeNO_x$ effect of the increasing amounts of unreacted NH_3. Here, the NO reduction reaction with NH_3 ($NH_3 + NO \rightarrow N_2 + H_2O + 0.5H_2$) is known to play a pivotal role, as NH_3 reacts with NO to form mostly N_2 and H_2O, thereby abating NO formed from other NH_3 consumption pathways. A 40% reduction in NO levels was attained, thus meeting the experimental trends from the literature, reporting a 30–50% NO emissions re-duction for coal–NH_3 co-firing in fluidized bed reactors operating at similar condi-tions [30, 32].

In Figure 6.2e, small concentrations of NH_3 emissions are predicted for 10% and 20% ratios (<5 ppm); as the NH_3 co-firing ratio increases, the amount of unreacted NH_3 at the reactor outlet gradually increases as well. Note that the lowest NO_x emis-sions are also attained at this relatively high NH_3 co-firing ratio. NH_3 slip is one of the main issues to be overcome when performing NH_3 combustion, and such levels must be monitored and reduced. Increasing NH_3 injection may benefit NO reduction, but it may also increase the risk of unburned NH_3; therefore, a balance in the NH_3/ NO ratio is required to enhance NH_3 reactivity and thus diminish NH_3 slip in the unburned streams [11].

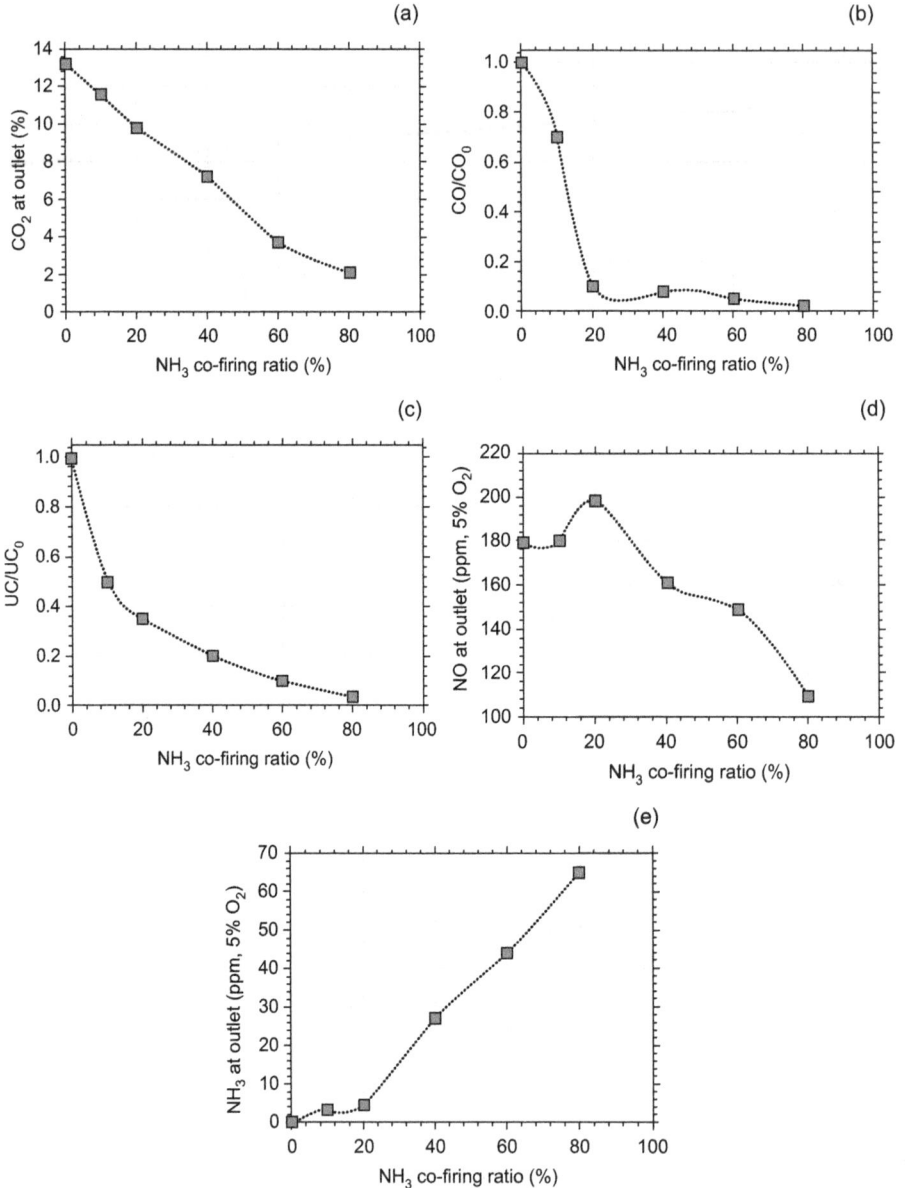

Figure 6.2: Effect of NH_3 co-firing ratio on numerical (a) CO_2, (b) normalized CO, (c) normalized UC, (d) NO, and (e) NH_3 emissions.

NH_3 is a colorless gas with a pungent odor whose odor detection concentration threshold is very low (around 1.5 ppm, or below), while the sensory concentration for irritation of eyes and airways ranges from 20 to 50 ppm [33]. NH_3 emission standards are extremely variable. Based on coal-fired systems operating in Europe and Japan, NH_3 emissions were limited at 5 ppm, although levels of 5–20 ppm are deemed acceptable, depending on its whereabouts [34]. Other reports state an average of NH_3 emission factors for coal fluidized bed reactors coupled with selective noncatalytic reduction systems around 20–30 ppm [35]. Currently, there is still limited regulation on NH_3 emissions as it is not a regulated pollutant (both in the United States and European Union); instead, it is regarded as a $PM_{2.5}$ precursor once it contributes significantly to their formation [36].

Nevertheless, concerning these ranges, the NH_3 emission limits for the co-firing in the studied fluidized bed reactor would be achievable up to a 20% ratio. At a 40% ratio operation, some emission restraints could be applied, depending on local environmental regulations and policies, while at 60–80% ratios additional abatement measures for NH_3 slip would be required for feasible operation.

In general, the numerical results attained for the NH_3 co-firing ratio effects on CO_2, CO, UC, NO, and NH_3 emissions in a coal-fired fluidized bed reactor are consistent with the trends found in the literature [1, 7, 8].

6.3.2 Coal–biomass–NH_3 co-firing: effect on the emissions reduction

The combined effect of biomass and NH_3 co-firing with coal maintained identical trends to the ones attained from coal–NH_3 co-firing, with CO_2, normalized CO, and NO gradually decreasing with the biomass–NH_3 ratio increase, while normalized NH_3 emissions increase (Figure 6.3a–d). The co-firing of coal and biomass, in particular, has already been long established and demonstrated as an effective measure to reduce GHG emissions and environmental pollution [23, 37, 38]. Here, biomass and NH_3 co-firing ratios were raised to 50% and 40% (by mass), respectively, while coal was gradually phased out from the mixture. Lessons learned from coal–NH_3 co-firing showed that 40% NH_3 ratio operation (<30 ppm) was possible regarding NH_3 emissions, although some emission restraints could be applied depending on local environmental regulations and policies [34]. Whereas at 60–80% ratios, substantial abatement measures for NH_3 emissions would be required for feasible operation. Thus, given these assumptions, a maximum of 40% NH_3 ratio was kept in the fuel mix.

Both biomass and coal are carbon-bearing fuels, yet CO_2 emissions are directly related to the fuel's carbon content [39]. The bituminous coal used in this work has higher fixed carbon and carbon elements than forest residues. Hence, fossil CO_2 emissions were continuously reduced by displacing coal and increasing the share of biomass and carbon-free NH_3 (Figure 6.3a). The same reasoning applies to CO emissions, as they decrease with both coal feed reduction and the increase in biomass and NH_3 ratio (Figure 6.3b) [13]. Ultimately, biomass is regarded as a carbon-neutral resource; therefore, CO_2 emission should decline proportionally to the amount of coal offset by biomass [40].

Apart from the previously discussed effect of NH_3 on NO abatement during coal co-firing, several advantages are addressed to biomass in reducing NO emissions (Figure 6.3c). Biomass contains lower nitrogen content than coal leading to reduce NO emissions. Plus, during combustion, most of the fuel nitrogen in biomass is converted to NH radicals (mainly NH_3); on the other hand, it reduces NO to N_2, hence lowering NO emissions [41]. Also, both biomass and NH_3 are known for decreasing peak flame temperature within the reactor [42, 43]. Biomass's lower calorific value and NH_3's low flammability and reluctance to combustion contribute to NO emission abatement, as higher temperatures promote NO formation [44]. This situation occurs as lower flame temperature leads to more char being generated. NO formation is reduced by char. Thus, as the flame temperature decreases, NO reduction by the char increases, leading to lower NO emissions [8]. Lastly, the high volatile content of biomass can also be used as a reburn fuel for NO reduction, providing biomass with additional potential in NO emissions abatement from coal combustion [41].

Similar to coal–NH_3 co-firing, NH_3 emissions showed to build up with the biomass–NH_3 ratio increase. As aforementioned, this increase is mainly attributed to the NH_3 ratio increase in the fuel mix. Most of the nitrogen in biomass is converted to NH_3 products during combustion, yet NH_3 destruction occurs to be converted into N_2 and H_2 [45]. Given the numerical results, the analyzed fluidized bed reactor could operate with coal–biomass–NH_3 up to 20% NH_3 with no NH_3 slip.

6.4 Final remarks

This numerical study showed that through biomass and NH_3 co-firing in a pilot-scale fluidized bed reactor, one can reduce carbon and NO emissions from coal combustion. The 2D Eulerian–Lagrangian model was validated against coal and biomass combustion experimental runs for different gas concentrations at several positions within the reactor with very reasonable agreement. The results showed that with the NH_3 co-firing increase, CO_2 emissions were reduced up to 26% compared to pure coal firing. CO emissions decreased with the H_2O concentration increase in the flue gas due to NH_3 combustion leading CO to react with H_2O, promoting CO abatement. UC decreased with NH_3 co-firing, in which both sidewall NH_3 injection and air-rich operating conditions were considered to lower UC concentrations. NO emissions remained the same level as pure coal combustion for a 10% ratio, while at 20–80% ratios, NO emissions decreased up to 40% due to the $DeNO_x$ effect and the NH_3 catalytic effect by reducing NO to form N_2 and H_2O. Small concentrations of NH_3 emissions were predicted at 10–20% ratios, while at 40% co-firing and above, the amount of unreacted NH_3 increased.

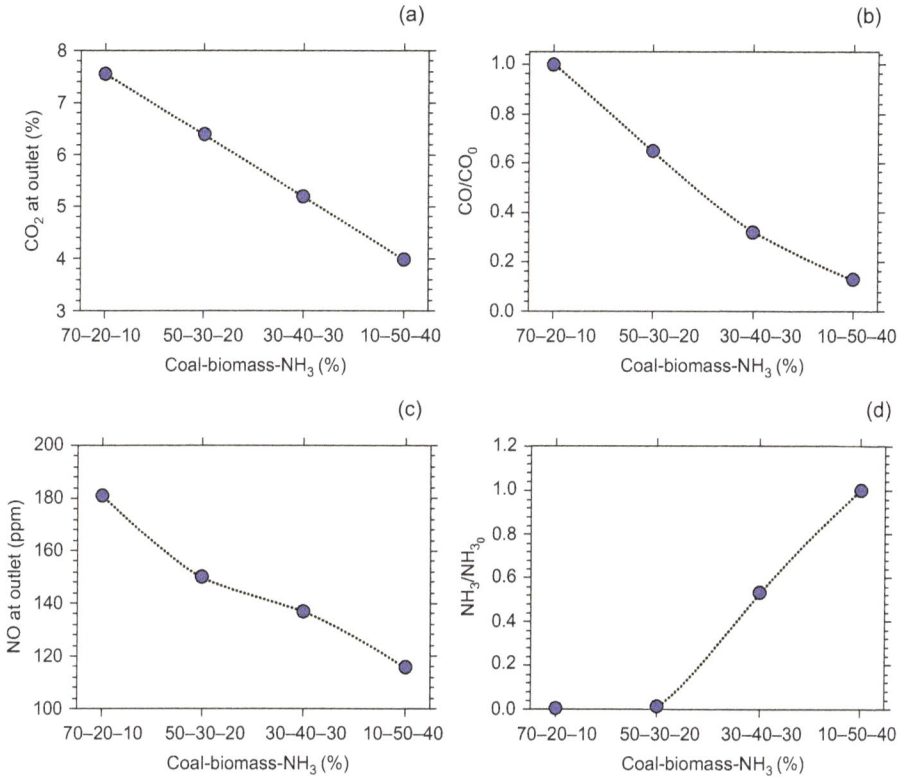

Figure 6.3: Effect of coal–biomass–NH$_3$ co-firing on (a) CO$_2$, (b) normalized CO, (c) NO, and (d) normalized NH$_3$ emissions.

Biomass, as a "carbon lean" fuel, comes as a valuable addition to this fuel mix providing supplementary feed resources while allowing locally sourced fuel options for coal-fired systems to avoid supply shortages.

Overall, the obtained results reinforce existing evidence for the use of NH$_3$ as a viable energy carrier for decarbonizing energy-intensive sectors. Also, this numerical study reinforces previous empirical demonstrations sustaining that the co-firing of NH$_3$ in coal-fired units is achievable with minor modifications to the existing systems within a 20% NH$_3$ range. Still, further research is required to characterize optimal NH$_3$ co-firing operation, namely the NH$_3$ effects on the systems' scale-up and efficiency, NH$_3$'s low calorific value, NH$_3$ slip issues, and NH$_3$ effects on volatilization products.

References

[1]   J. Zhang, T. Ito, H. Ishii, S. Ishihara, and T. Fujimori, "Numerical investigation on ammonia co-firing in a pulverized coal combustion facility: Effect of ammonia co-firing ratio," *Fuel*, vol. 267, p. 117166, 2020.

[2]   O. Elishav, B. M. Lis, E. M. Miller, D. J. Arent, A. Valera-Medina, A. G. Dana, et al., "Progress and prospective of nitrogen-based alternative fuels." *Chemical Reviews*, vol. 120, pp. 5352–5436, 2020.

[3]   Z. Wan, Y. Tao, J. Shao, Y. Zhang, and H. You, "Ammonia as an effective hydrogen carrier and a clean fuel for solid oxide fuel cells," *Energy Conversion and Management*, vol. 228, p. 113729, 2021.

[4]   D. Flórez-Orrego, F. Maréchal, and S. O. Junior, "Comparative exergy and economic assessment of fossil and biomass-based routes for ammonia production," *Energy Conversion and Management*, vol. 194, pp. 22–36, 2019.

[5]   A. Valera-Medina, H. Xiao, M. Owen-Jones, W. I. F. David, and P. J. Bowen, "Ammonia for power," *Progress in Energy and Combustion Science*, vol. 69, pp. 63–102, 2018.

[6]   J. S. Cardoso, V. Silva, R. C. Rocha, M. J. Hall, M. Costa, and D. Eusébio, "Ammonia as an energy vector: Current and future prospects for low-carbon fuel applications in internal combustion engines," *Journal of Cleaner Production*, vol. 296, p. 126562, 2021.

[7]   M. Tamura, T. Gotou, H. Ishii, and D. Riechelmann, "Experimental investigation of ammonia combustion in a bench scale 1.2 MW-thermal pulverised coal firing furnace," *Applied Energy*, vol. 277, p. 115580, 2020.

[8]   S. Ishihara, J. Zhang, and T. Ito, "Numerical calculation with detailed chemistry on ammonia co-firing in a coal-fired boiler: Effect of ammonia co-firing ratio on NO emissions," *Fuel*, vol. 274, pp. 117742, 2020.

[9]   R. C. Rocha, M. Costa, and X. S. Bai, "Chemical kinetic modelling of ammonia/hydrogen/air ignition, premixed flame propagation and NO emission," *Fuel*, vol. 246, pp. 24–33, 2019.

[10] IHI. IHI and Partners Launching Ammonia Co-Firing Technology Feasibility Assessments at Coal Power Stations in Malaysia with Partners and for Other Companies to Establish Local Carbon-Free Ammonia Supply Chain, 2021. Available from: https://www.ihi.co.jp/en/all_news/2021/resources_energy_environment/1197552_3360.html. [Accessed January 30 2022].

[11] A. Valera-Medina, F. Amer-Hatem, A. K. Azad, I. C. Dedoussi, M. Joannon, R. X. Fernandes, et al., "Review on ammonia as a potential fuel: From synthesis to economics." *Energy & Fuels*, vol. 35, pp. 6964–7029, 2021.

[12] Ammonia Energy News, Chugoku Electric Completes Successful Trial, Seeks Patent for Ammonia Co-Firing Technology, 2017. Available from: https://www.ammoniaenergy.org/articles/chugoku-electric-completes-successful-trial-seeks-patent-for-ammonia-co-firing-technology/. [Accessed November 25 2021].

[13] S. H. Crolius, IHI First to Reach 20% Ammonia-Coal Co-Firing Milestone, 2018. Available from: https://www.ammoniaenergy.org/articles/ihi-first-to-reach-20-ammonia-coal-co-firing-milestone/. [Accessed March 22 2021].

[14] L. A. C. Tarelho, M. A. A. Matos, and F. J. M. A. Pereira, "Axial and radial CO concentration profiles in an atmospheric bubbling FB combustor," *Fuel*, vol. 84, pp. 1128–1135, 2005.

[15] L. A. C. Tarelho, D. S. F. Neves, and M. A. A. Matos, "Forest biomass waste combustion in a pilot-scale bubbling fluidised bed combustor," *Biomass and Bioenergy*, vol. 35, pp. 1511–1523, 2011.

[16] J. Xie, W. Zhong, B. Jin, Y. Shao, and H. Liu, "Simulation on gasification of forestry residues in fluidized beds by Eulerian–Lagrangian approach," *Bioresource Technology*, vol. 121, pp. 36–46, 2012.

[17]   ANSYS, ANSYS Fluent Theory Guide. Release 15.0. ANSYS, Inc., 2013.

[18]   X. Ku, T. Li, and L. T. Eulerian–Lagrangian, "Simulation of biomass gasification behavior in a high-temperature entrained-flow reactor," *Energy & Fuels*, vol. 28, pp. 5184–5196, 2014.

[19]   B. Launder and B. Spalding, *Lectures in Mathematical Models of Turbulence*, London, England: Academic Press, 1972.

[20]   L. Yu, J. Lu, X. Zhang, and S. Zhang, "Numerical simulation of the bubbling fluidized bed coal gasification by the kinetic theory of granular flow (KTGF)," *Fuel*, vol. 86, pp. 722–734, 2007.

[21]   S. S. Park, H. J. Jeong, and J. Hwang, "3-D CFD modeling for parametric study in a 300-MWe one-stage oxygen-blown entrained-bed coal gasifier," *Energies*, vol. 8, pp. 4216–4236, 2015.

[22]   L. D. Smoot and P. J. Smith, *Coal Combustion and Gasification*, New York, USA: Plenum Press, 1985.

[23]   C. Chen, X. Wu, and L. Zhao, "Simulation of coal and biomass cofiring with different particle density and diameter in bubbling fluidized Bed under O2/CO2 atmospheres," *Journal of Combustion*, vol. 2018, p. 6931483, 2018.

[24]   J. P. Kim, U. Schnell, and G. Scheffknecht, "Comparison of different global reaction mechanisms for MILD combustion of natural gas," *Combustion Science and Technology*, vol. 180, pp. 565–592, 2008.

[25]   J. B. Howard, G. C. Williams, and D. H. Fine, "Kinetics of carbon monoxide oxidation in postflame gases," *Symposium (International) on Combustion*, vol. 14, pp. 975–986, 1973.

[26]   J. Brouwer, M. P. Heap, D. W. Pershing, and P. J. Smith, "A model for prediction of selective noncatalytic reduction of nitrogen oxides by ammonia, urea, and cyanuric acid with mixing limitations in the presence of co," *Symposium (International) on Combustion*, vol. 26, pp. 2117–2124, 1996.

[27]   W. D. Monnery, K. A. Hawboldt, A. E. Pollock, and W. Y. Svrcek, "Ammonia pyrolysis and oxidation in the Claus furnace," *Industrial & Engineering Chemistry Research*, vol. 40, pp. 144–151, 2001.

[28]   C. Chen, M. Horio, and T. Kojima, "Numerical simulation of entrained flow coal gasifiers. Part I: Modeling of coal gasification in an entrained flow gasifier," *Chemical Engineering Science*, vol. 55, pp. 3861–3874, 2000.

[29]   H. Freund, "Gasification of carbon by CO2: A transient kinetics experiment," *Fuel*, vol. 65, pp. 63–66, 1986.

[30]   E. Hampartsoumian and M. Gibbs, The influence of NH3 addition on the NO emissions from a coal fired fluidised bed combustor. 19th Symposium (International) on Combustion/The Combustion Institute, pp. 1253–1262, 1982.

[31]   A. Yamamoto, M. Kimoto, Y. Ozawa, and S. Hara, Basic co-firing characteristics of ammonia with pulverized coal in a single burner test furnace. 2018 NH3 Fuel Conference, 2018.

[32]   B. M. Gibbs, T. F. Salam, S. F. Sibtain, R. J. Pragnell, and D. W. Gauld, The reduction of NOx emissions from a fluidized bed combustor by staged combustion combined with ammonia addition 22nd Symposium (International) on Combustion/The Combustion Institute, 1988, pp. 1147–1154.

[33]   M. Li, C. J. Weschler, G. Bekö, P. Wargocki, G. Lucic, and J. Williams, "Human ammonia emission rates under various indoor environmental conditions," *Environmental Science & Technology*, vol. 54, pp. 5419–5428, 2020.

[34]   Office of Air Quality Planning and Standards, Control and Pollution Prevention Options for Ammonia Emissions, U.S. Environmental Protection Agency, 1995.

[35]   Atmospheric Research and Exposure Assessment Laboratory, *Development and Selection of Ammonia Emission Factors*, Washington, D.C.: U.S. Environmental Protection Agency, 1994.

[36] E. Giannakis, J. Kushta, A. Bruggeman, and J. Lelieveld, "Costs and benefits of agricultural ammonia emission abatement options for compliance with European air quality regulations," *Environmental Sciences Journal*, vol. 31, pp. 93–106, 2019.

[37] G. T. Marangwanda, D. M. Madyira, H. C. Chihobo, and T. O. Babarinde, "Modelling co-combustion of bituminous coal and pine sawdust: Thermal behaviour," *Fuel Communications*, vol. 9, p. 100035, 2021.

[38] Q. N. Hoang, M. Vanierschot, J. Blondeau, T. Croymans, R. Pittoors, and J. Van Caneghem, "Review of numerical studies on thermal treatment of municipal solid waste in packed bed combustion," *Fuel Communications*, vol. 7, p. 100013, 2021.

[39] EPA, Greenhouse Gas Inventory Guidance, Direct Emissions from Stationary Combustion Sources U.S. Environmental Protection Agency, 2016.

[40] J. M. Jones, A. B. Ross, E. J. S. Mitchell, A. R. Lea-Langton, A. Williams, and K. D. Bartle, "Organic carbon emissions from the co-firing of coal and wood in a fixed bed combustor," *Fuel*, vol. 195, pp. 226–231, 2017.

[41] S. G. Sahu, N. Chakraborty, and P. Sarkar, "Coal–biomass co-combustion: An overview," *Renewable and Sustainable Energy Reviews*, vol. 39, pp. 575–586, 2014.

[42] A. A. Bhuiyan and J. Naser, "CFD modelling of co-firing of biomass with coal under oxy-fuel combustion in a large scale power plant," *Fuel*, vol. 159, pp. 150–168, 2015.

[43] H. Kobayashi, A. Hayakawa, K. D. K. A. Somarathne, and E. C. Okafor, "Science and technology of ammonia combustion," *Proceedings of the Combustion Institute*, vol. 37, pp. 109–133, 2019.

[44] T. Cai, D. Zhao, B. Wang, J. Li, and Y. Guan, "NOx emission and thermal performances studies on premixed ammonia-oxygen combustion in a CO2-free micro-planar combustor," *Fuel*, vol. 280, pp. 118554, 2020.

[45] A. H. Tchapda and S. V. Pisupati, "A review of thermal co-conversion of coal and biomass/waste," *Energies*, vol. 7, no. 3, pp. 1098–1148, 2014.

Future prospects in gasification

There is no doubt that if the aim is to reduce reliance on fossil fuels and CO_2 emissions, gasification processes come to the fore. Biomass-to-energy conversion using gasification to produce syngas, heat, hydrogen, ethanol, and power has proven successful. However, some critics remain skeptical about the commercial future of gasification given its installation costs, unproven technologies, and lower power yields; nevertheless, the potential for resource output is still far greater.

One current and highly relevant use for gasification processes is its ability for supporting the future of green hydrogen and green ammonia production, to which biomass gasification has been promoted as a promising method due to its efficiency in dealing with these processes, environmental performance, hydrogen-rich syngas, and high flexibility in dealing with alternative feedstocks. In certain application scenarios, biomass gasification does stand out as a cost-effective and environmentally effective energy source, avoiding the need for breakthrough advances on catalysts while delivering clear feedstock cost benefits and favorable economies-of-scale effects [1].

Furthermore, developments in gasification technologies still advance significantly with some tremendous developments in gasification systems such as plasma gasification, supercritical water gasification, co-gasification, tar and soot elimination, integrated gasification systems, and a combination of gasification and anaerobic digestion, producing promising findings that could further enhance gasification systems in the coming decades [2].

The development of gasification systems has also included extensive research on product gas cleanup. Raw product gases may contain particulates, tars, ammonia, and other impurities that can interfere with downstream processes and components or create emission problems. It is therefore crucial to understand optimal gasification process parameters for practical design and operation for maximizing the potential. Process optimization of biomass gasification allows reduction of the energy loss caused by pretreatment of the biomass prior to the conversion process, optimizing the carbon conversion efficiency in the reactor, reducing tar production, and cleaning the syngas for further processing [3].

Gasification is a waste-to-energy conversion scheme that offers a most attractive solution to both waste disposal and energy problems. However, gasification still has some economic and technical challenges, concerning the nature of the solid waste residues and its heterogeneity. When dealing with MSW, the greatest strength of gasification lies in its environmental performance, since emission tests indicate that gasification meets the existing limits, and it can also have an important role in the reduction of landfill disposal [4]. Overall, MSW gasification has been attempted at various scales for many decades but has yet to become a fully mature technology. A key challenge is still to demonstrate the scale and robustness required for the gasification of mixed MSW.

https://doi.org/10.1515/9783110758214-007

Although commercial-scale gasification systems (>1 MW) are usually preferable by developed nations, complete usage of biomass energy in remote and local regions would occur through the utilization of small-scale gasifiers (<200 kW). Consequently, this increased the development of small-scale gasification systems across the world. In addition to its enhanced feasibility and cost-effectiveness compared to commercial systems, small-scale units are pivotal for off-grid energy solutions providing energy access to decentralized areas and/or rural household communities (particularly in developing countries), granting alternative electric power solutions to communities where connection to the central grid is economically unfeasible [5]. This feature reinforces gasification's versatility setting it as an important asset to the renewable energy portfolio.

Bioenergy and advanced biofuels from biomass and wastes produced through gasification will end up playing a valuable role in the global sustainable energy system. Currently, there is a set of relevant low-carbon technologies with light-emitting diodes, solar photovoltaic, onshore wind, and batteries powering hybrid and electric vehicles as the front-runners [6]. The bioenergy sector, on the other hand, seems to be going through a transitional phase with strong oscillations derived from changing political policies [7]. In fact, the pace of these technologies depends not only on how they shape their performance and cost but also on the regulatory measures from national and international governments and organizations. Therefore, bioenergy seems to lose its strength in recent times due to changing policies; the absence of a regulated market for biomass pricing, the lack of standardized pretreatment procedures that can provide a consistent product, and some of the technologies related to biomass conversion are still characterized by obsolete procedures regarding energy efficiency. Notwithstanding, bioenergy has a very promising future as only a very small fraction of its potential has been exploited so far [8]. That potential can easily be achieved given the huge quantities of unused bioresidues around the world that can be converted into bioenergy. Thus, bioenergy and advanced biofuels from biomass and wastes produced through gasification will end up playing a valuable role in the global sustainable energy system, setting gasification as a highly valuable key enabler to decarbonizing energy consumption.

References

[1] J. S. Cardoso, V. Silva, J. A. M. Chavando, D. Eusébio, M. J. Hall, and M. Costa, "Small-scale biomass gasification for green ammonia production in Portugal: A techno-economic study," *Energy & Fuels*, vol. 35, no. 17, pp. 13847–13862, Sep. 2021, doi: 10.1021/ACS. ENERGYFUELS.1C01928.

[2] A. Ramos, E. Monteiro, V. Silva, and A. Rouboa, "Co-gasification and recent developments on waste-to-energy conversion: A review," *Renewable and Sustainable Energy Reviews*, vol. 81, Elsevier Ltd, pp. 380–398, 2018, doi: 10.1016/j.rser.2017.07.025.

[3]   Z. Barahmand and M. S. Eikeland, "A scoping review on environmental, economic, and social impacts of the gasification processes," *Environments*, vol. 9, no. 7, p. 92, Jul. 2022, doi: 10.3390/ENVIRONMENTS9070092.

[4]   N. Couto, et al., "Numerical and experimental analysis of municipal solid wastes gasification process," *Applied Thermal Engineering*, vol. 78, pp. 185–195, Mar. 2015, doi: 10.1016/J. APPLTHERMALENG.2014.12.036

[5]   J. Sousa Cardoso, V. Silva, D. Eusébio, I. Lima Azevedo, and L. A. C. Tarelho, "Techno-economic analysis of forest biomass blends gasification for small-scale power production facilities in the Azores," *Fuel*, vol. 279, pp. 118552, Nov. 2020, doi: 10.1016/J. FUEL.2020.118552.

[6]   J. Cardoso, V. B. R. e Silva, D. Eusébio, J. Cardoso, V. B. R. e Silva, and D. Eusébio, "Introductory chapter: Low carbon economy. An overview," *Low Carbon Transition – Technical, Economic and Policy Assessment*, Oct. 2018, doi: 10.5772/INTECHOPEN.80920.

[7]   A. Ramos, E. Monteiro, V. Silva, and A. Rouboa, "Co-gasification and recent developments on waste-to-energy conversion: A review," *Renewable and Sustainable Energy Reviews*, vol. 81, pp. 380–398, 2018, doi: 10.1016/j.rser.2017.07.025.

[8]   UN and GEF, "The role of bioenergy in the clean energy transition and sustainable development lessons from developing countries," 2021.

Index

https://doi.org/10.1515/9783110758214-008

www.ingramcontent.com/pod-product-compliance
Lightning Source LLC
Chambersburg PA
CBHW081540220326
41598CB00036B/6505